THE STUDY OF FOSSILS

HUTCHINSON BIOLOGICAL MONOGRAPHS
GENERAL EDITOR: K. Simkiss

The Biology of Fungi	C. T. Ingold
Cells and Cell Structure	E. H. Mercer
Bird Flight	K. Simkiss
The Study of Fossils	J. F. Kirkaldy
Essays on Protozoology	H. Sandon
Marine Plankton	G. E. and R. C. Newell
Bird Structure	D. A. Ede
A Biology of Marine Algae	A. D. Boney
A Guide to Invertebrate Fossils	F. A. Middlemiss
Enzymes, their Nature and Role	A. Wiseman and B. J. Gould
Environmental Pollution by Chemicals	Colin Walker

THE STUDY OF FOSSILS

J. F. KIRKALDY

Professor of Geology
Queen Mary College, University of London

HUTCHINSON EDUCATIONAL

HUTCHINSON EDUCATIONAL LTD
178–202 Great Portland Street, London W1

London Melbourne Sydney
Auckland Johannesburg Cape Town
and agencies throughout the world

First published 1963
Second impression 1964
Revised edition June 1971

© J. F. Kirkaldy 1963

Printed in Great Britain by litho on smooth wove paper
by Anchor Press, and bound by Wm. Brendon,
both of Tiptree, Essex
ISBN 0 09 108391 5

Contents

	Preface	9
I	THE ROCKS THAT CONTAIN FOSSILS	11
II	THE STRATIGRAPHICAL TABLE AND THE ABSOLUTE DATING OF ROCKS	25
III	THE NATURE AND PRESERVATION OF FOSSILS	29
IV	THE SUCCESSION OF LIFE	40
V	THE USE OF FOSSILS AS TIME-MARKERS	65
VI	FOSSILS AS INDICATORS OF ENVIRONMENTAL CONDITIONS	78
VII	PALAEOCLIMATES	88
VIII	SPECIES IN FOSSILS AND EVOLUTION	95
	Appendix: The Chief Divisions of the Geological Column	106
	Further Reading	109
	Glossary	111
	Index	113

Figures

1	Stratification at Lyme Regis, Dorset	13
2	Fossil tree in position of life, Victoria Park, Glasgow	14
3	The double unconformity at Loch Assynt, Sutherland	15
4	A block of limestone cut by a non-sequence	16
5	A sandpit showing true- and false-bedding	17
6	Measurement of dip and strike with a compass-clinometer	17
7	Chief types of folds	18
8	Structures produced by faulting	19
9	Thin-sections of limestones	22
10	An oil-well drilling rig	34
11	Examples of microfossils	35
12	An archaeocyathid (after Okulitch)	36
13	The effect of cleavage	37
14	The structure of dendroid graptolites (after Bulman)	38
15	'Fossils' from the Pre-Cambrian rocks	41
16	Some Cambrian fossils	43
17	Graptolites of the Ordovician rocks	44
18	Fossils of the shelly facies of the Ordovician	45
19	Thecal variation in the monograptids (after Bulman)	45
20	Ballstones in the Wenlock Limestone	46
21	Fossils of the shelly facies of the Silurian	47
22	Ostracoderms and placoderms	48
23	Sharks and bony fish	49
24	Devonian and Carboniferous plants	50
25	Typical fossils of the Carboniferous Limestone	51

FIGURES

26	Fossils of the Millstone Grit and the Coal Measures	52
27	A Coal Measure cyclothem	53
28	Fossils from the marine Permian and Triassic rocks	54
29	Fossils of the Karroo System	55
30	Typical fossils of the Jurassic clays	56
31	Typical fossils of the Jurassic limestones	57
32	Mesozoic plants	58
33	Marine fossils of the Cretaceous System	60
34	Jurassic and Cretaceous reptiles	61
35	Some Tertiary mammals	61
36	Fossils of the English Tertiary Beds	63
37	Evolutionary changes in *Micraster*	69
38	Relations of the Upper Greensand and the Gault	71
39	Two interpretations of the structure of the Southern Uplands of Scotland	74
40	S. S. Buckman's work in the Cotswolds	75
41	Some of the structures in which oil may accumulate	76
42	Palaeoecology of a two-foot band of shale (after Craig)	79
43	A bedding plane showing current-aligned monograptids	81
44	A Jurassic reef-belt (after Arkell)	84
45	The Capitan reef-belt (after Newell)	85
46	Changing communities during the upward growth of a Silurian reef (after Lowenstam)	86
47	Change of sea temperatures since Palaeocene times along the west coast of North America (after Durham)	91
48	Climatic provinces shown by the distribution of Jurassic ammonites (after Arkell)	92
49	Temperature variation during the Cretaceous Period (after Bowen)	93
50	The distribution of recent and fossil coral reefs (after Termier and Termier)	94
51	Members of the *Ostrea-Gryphea* lineage (after Trueman)	97
52	Variation curves of oyster populations at different stratigraphical levels	97
53	Variation in a community of Coal Measure lamellibranchs (after Leitch)	98
54	Changes in the forefeet of the horse series (after Simpson)	100
55	Changes in the brains of the horse series (after Simpson)	101
56	Evolution of the horses (after Simpson)	102
57	Horizontal variation in brachiopods of the Fuller's Earth Rock (after McKerrow)	103
58	Faunal provinces and migration routes of Middle Liassic (Domerian) brachiopods (after Ager)	104

Preface

This little book has been written at the suggestion of the General Editor, who thought it would be valuable to biologists to know the type of information geologists hope to obtain from the study of fossils. In the first three chapters I have attempted to give, in I hope not too succinct a form, the necessary geological background. Then in Chapter IV the succession of life since at least Pre-Cambrian times is outlined. Limitation of space prohibits any detail of the morphology of the various groups mentioned, but each one is shown on at least one figure and references are given at the end for further reading. The last four chapters deal with the various ways in which a geologist may make use of palaeontological information when he is attempting to unravel the stratigraphical history and structure of an area. It was thought most helpful to cite a number of examples and to discuss these at some length. Partly owing to pressure of space and partly because the author is primarily a stratigrapher, rather than a systematic palaeontologist, evolutionary aspects have not been discussed in detail.

The great majority of figures were drawn by Dr J. Machin, to whom I should like to acknowledge my grateful thanks for the care he has taken. Certain figures have appeared in my *General Principles of Geology*, published by Messrs Hutchinson, and others are based on G. G. Simpson's *Horses*, published by Oxford University Press. I should like to thank my colleagues, Professor G. E. Newell and Dr F. A. Middlemiss, for reading and improving the typescript, the one from a zoological, the other from a geological, point of view, and the publishers for the care they have taken.

J. F. KIRKALDY

Queen Mary College, 1963

Words explained in the Glossary (p. 111) are italicized on their first appearance in the text.

I
The Rocks that contain Fossils

IGNEOUS, sedimentary and metamorphic are the three main classes of rocks. Igneous rocks have solidified from molten material. They are composed of tightly interlocking minerals. According to their cooling history, igneous rocks may be coarse-grained, as in granite, with the different minerals easily recognizable, or they may be finer-textured, as in basalt, with scattered minerals just perceptible to the naked eye, set in a fine-grained ground-mass whose crystalline nature can be recognized only in thin-section under the petrological microscope. The molten material (magma) may have been chilled so quickly as to be glasslike (obsidian), or as in pumice it may be full of the cavities blown by hot gases escaping from the top of a lava flow. Such rocks clearly are extremely unlikely to contain fossils, the remains of the life of the past.

Sedimentary rocks—sandstone, clay, limestone, coal, etc.—have been formed from the erosion of pre-existing rocks. Rocks under the attack of rain, ice, waves, wind, and the stresses set up by considerable diurnal change of temperature, break down into solid fragments, whilst the more easily soluble constituents of the rocks, if in contact for sufficient time with water, pass into solution. The solid particles are transported under gravity by moving water (including ice) or by the wind, to be deposited in the seas or on the land surfaces to form new sedimentary rocks. The dissolved salts may be precipitated to form beds of rock salt, gypsum, etc., or they may be absorbed by organisms to form their skeletons. This skeletal material may accumulate in sufficient abundance to build up distinctive varieties of sedimentary rocks, particularly limestone, or it may form but a minor constitutent of a sedimentary rock.

Metamorphic rocks are igneous or sedimentary rocks which have been altered by high temperatures, high pressures or a combination of these. When molten magma at a temperature of 600 or more degrees Centigrade is extruded as a lava flow, or intruded as a dyke, sill or granite *batholith*, the '*country rock*' with which it comes in contact is baked and its mineral composition changed. When portions of the Earth's crust are compressed during folding movements, particularly during periods of orogenesis (mountain building), the rocks involved will be greatly altered. They may suffer plastic deformation and considerable crystallization as a result of the formation of new minerals of high density, whilst the folded rocks may be forced down into regions of higher temperature. Later the folded mountain chain will be uplifted and eroded. Under conditions of strong metamorphism any fossils in sedimentary rocks will be obliterated, but if the metamorphism has been not too great the more resistant fossils may not be completely destroyed. For example, fractured and elongated *guards* of belemnites have been found in the mica-schists of the Swiss Alps, much distorted trilobites in the famous roofing slates of Llanberis in North Wales, vestiges of corals in limestone bands amongst the strongly deformed rocks of the Bergen district in Norway. But such finds are exceptional, and in general the hope of finding fossils in metamorphic rocks is slight.

Stratification

Sedimentary rocks are usually stratified; that is, they occur in layers. The strata or layers may be sharply defined as at Lyme Regis, Dorset. There the cliffs and foreshore are made up of layers or beds a few inches in thickness, of pale blue-grey clayey limestone separated by similar thicknesses of darker-coloured clay rocks (shales or mudstones) (Fig. 1). In the chalk cliffs of Beachy Head, Eastbourne, or around Margate and Ramsgate in east Kent, the stratification is picked out by lines of flint or by thin seams of chalk, which on careful inspection differ slightly in colour and texture from the beds above and below. The lines of stratification in rocks are bedding planes, marking the position for a brief period of time of a former sea floor, lake floor, delta plain, desert swamp or other land surface. On the limestone bedding planes or 'pavements' at Lyme Regis can be seen the cross-sections of many ammonites, the larger ones several feet in diameter. Ammonites were nektonic (swimming) cephalopods, whose shells after death settled to lie horizontally on the sea floor. A coal seam, composed of closely packed plant debris, is not only bedded but beneath it is often a fire-clay or *ganister* sandstone, the fossil soil on which the plants grew, with the roots of plants spreading out along the bedding and the stumps of trees at right angles to the bedding (Fig. 2). The 'roof' of a coal seam is often a thin layer of shale in which occur marine fossils, such as brachiopods, goniatites (another type of extinct cephalopods) and crinoids. An advance of the sea has submerged the swamps, killing off the luxuriant vegetation which was subsequently compacted to form coal.

Unconformity and non-sequence

Through thicknesses of rocks, measurable perhaps in thousands of feet, the bedding planes may be sensibly parallel to one another. Such a succession of conformable beds must have been laid down during a period of time when the floor of the basin of deposition was steadily subsiding. But strata are not always conformable. At Vallis Vale, near Frome in Somerset, a line of quarries expose honey-coloured rather friable limestones resting horizontally on dark grey limestones, much harder and with their bedding planes not horizontal, but inclined or dipping at about 45 degrees. One can walk over the upper surface of the lower limestones. It is coated with oysters and bored by various

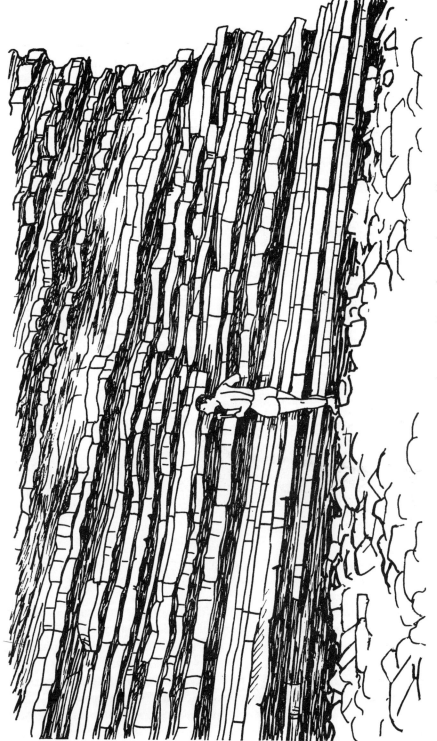

Fig. 1. Stratification at Lyme Regis, Dorset (drawn from a photograph)

marine organisms; the borings are infilled by material identical with the soft yellow limestone. In the basal few inches of the upper limestone are pebbles of the lower limestone. Both limestones are fossiliferous, but yield entirely different assemblages of fossils. The lower limestones must have been laid down horizontally in a sea, for they yield marine fossils. Then they were uplifted, tilted and eroded. After a long period of

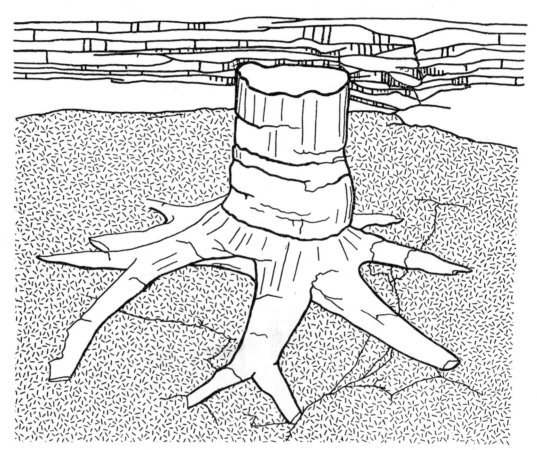

FIG. 2. Fossil tree in the position of life, Victoria Park, Glasgow. The outcrop in the background is of Upper Carboniferous rocks (drawn from a photograph)

time, as indicated by the complete change in fossil content, they sank beneath the sea again to form the 'hard ground' across which the upper limestones were deposited.

The upper limestones overlie the lower ones unconformably. The surface of unconformity separating the two groups is in this case a plane, cut by waves, but in other instances of unconformity the younger beds may rest on an irregular surface, a fossil land surface with hills and valleys, cut across the older beds (Fig. 3).

The quarries at Vallis Vale also demonstrate an important principle, the Law of Superposition; that is, a younger layer rests on or overlies an older bed. By following this law, beds can be arranged in order of their geological age. But as we shall see later there are exceptions to it.

Even in a conformable succession of beds there may be evidence that deposition was not a continuous process. In some of the limestone quarries of the Cotswolds may be found bedding planes coated with oysters which grew across a bored surface of the

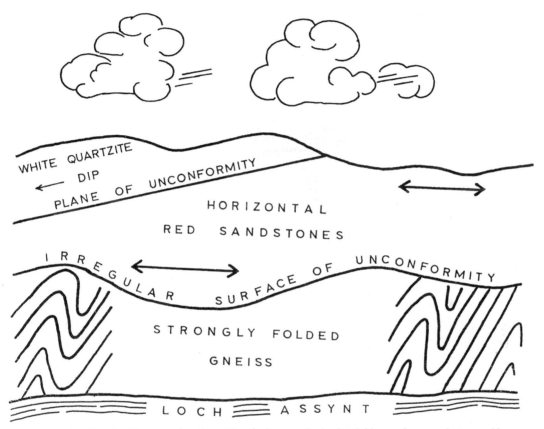

FIG. 3. The double unconformity at Loch Assynt, Sutherland (drawn from a photograph)

underlying limestone. The beds overlying the layer of oysters may contain pebbles of the underlying limestone. Though the phenomena resemble those at Vallis Vale, there is here no difference in dip or obvious difference in the fossil content of the beds above and below the surface of discontinuity, which is therefore called a non-sequence (Fig. 4).

Dip and strike

Stratification may not always be so easy to detect as the examples above suggest. Limestone, in particular, may be massive, with clearly defined bedding planes widely separated. Sand and sandstone exposures may reveal beds which at first sight dip in a confusing variety of different directions. Such beds may have been built up as sand-dunes or as subaqueous banks under the influence of fluctuating wind or water currents. It is important to distinguish between this false-bedding, indicating the rapidly changing positions of built-up surfaces, and the true-bedding which reflects the position of the floor of the basin of deposition. In Fig. 5 some of the criteria for distinguishing true- from false-bedding are shown. The seams of clay and the pebbly layers lie along the planes of true-bedding.

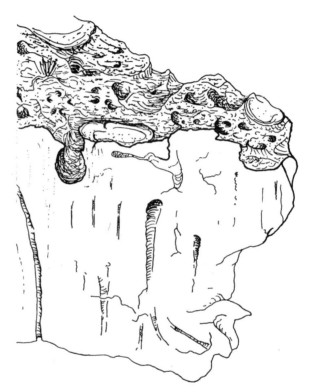

FIG. 4. A block of limestone cut by a non-sequence. The upper surface is coated with oysters and penetrated by borings, some of which are seen in section on the sides of the block

It is most important to determine true dip. In elucidating the geology of an area it is essential to know both the order of succession of the various rock layers and their disposition. The disposition of strata in space can be determined as shown on Fig. 6 by measuring their dip, that is the greatest angle made with the horizontal, and their direction of strike, that is the bearing of a horizontal line drawn on a bedding surface. These two measurements fix the position of the surface in three dimensions. A geological map shows where different beds, recognizable by their distinctive lithology (mass-character) and perhaps their fossil content, outcrop or cut the surface of the ground. The outcrop of a bed, which is lying horizontally, is parallel to the contours of the land surface, but if the bedding is dipping, then its outcrop will be elongated along the direction of strike. If the dip is gentle, 10 degrees to the horizontal or less, the outcrop will be sinuous, for it

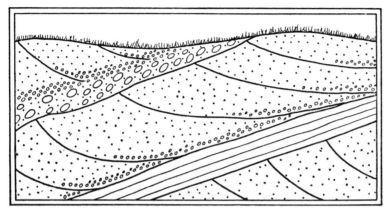

FIG. 5. Sketch of a sandpit showing true-bedding and false-bedding

is controlled more by the irregularities of the land surface than by the plane of the bed. But as the dip steepens the outcrop of the bed becomes straighter, until when the bed

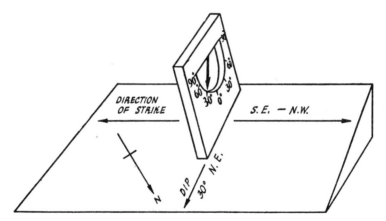

FIG. 6. Measurement of dip and strike with a compass-clinometer

is vertical (dipping at 90 degrees) its outcrop must cut straight across country without paying any attention to relief.

Folds and faults

If beds are dipping in a uniform direction, then their outcrops will be parallel, with the younger beds always occurring on the down-dip side. But if a map shows parallel or roughly parallel outcrops of the same bed, this bed must be folded for it to cut the

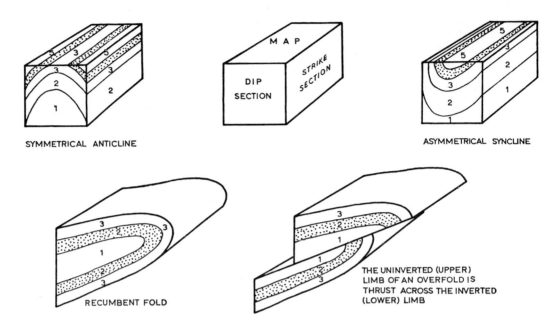

Fig. 7. The chief types of folds. The beds are numbered in order of deposition

ground surface twice. The chief types of folds are shown in Fig. 7. It will be noticed that the Law of Superposition does not hold on the lower or inverted limb of a recumbent fold or overfold. Recumbent folds are clearly the result of strong pressure and beds thus folded often show signs of metamorphism.

If the beds have been subjected to tension, they will fracture with the formation of faults (Fig. 8). The beds may also shear or fracture under compressive stress, producing tear faults with mainly horizontal movement of blocks relative to one another; or the pressure may be great enough for a recumbent fold to be so drawn out that finally the upper limb is thrust or driven across the inverted limb.

If folded or faulted beds are overlain unconformably by undisturbed beds, then clearly the tectonic disturbances must have occurred before the deposition of the unconformable upper layer.

Joints

Sedimentary rocks are often traversed by fractures, running at right angles to the bedding. Comparison of the sides of the fractures shows that there has been little or no relative movement along the joint. There are usually two sets of joints at right angles to each other, and if the rock is well jointed it breaks easily into rectangular blocks.

The chief types of Sedimentary rocks

On the basis of their physical and chemical characteristics, sedimentary rocks can be divided into the Clastic rocks, made up largely of fragments of older rocks, and the

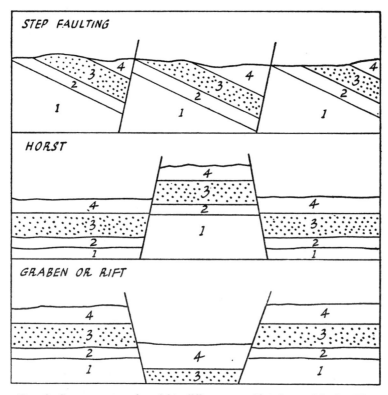

Fig. 8. Structures produced by different combinations of faults. The beds are numbered in order of deposition

Non-Clastic rocks, which are either wholly or partly of organic origin, or chemical precipitates. In addition, we must mention the Pyroclasts, material of volcanic origin deposited in a sedimentary environment. The rock-types which we shall be considering are listed below:

Clastic rocks	Rudaceous Rocks	Boulder and pebble beds
	Arenaceous Rocks	Sands
	Argillaceous Rocks	Clays
Non-Clastic rocks	Calcareous Rocks	Limestones
	Evaporite Deposits	Rock salt, gypsum, etc.
	Ferriferous Rocks	Iron ores of sedimentary origin
	Siliceous Deposits	Flint and chert
	Phosphatic Deposits	
	Carbonaceous Deposits	Peat, lignite, coal and oil accumulations
Pyroclasts		Ashes and agglomerates

Except for the evaporite deposits, the members of all groups could have been formed in a wide variety of environments. Indeed, it is often difficult to infer the precise environment in which any particular sedimentary rock was deposited.

It must also be clearly realized that the dividing lines between these groups are entirely arbitrary and that each group includes a large variety of rock-types which grade into one another and into those of other groups. A typical limestone is easily distinguishable from a typical sandstone; but one sometimes finds a rock consisting of sand grains set in a calcareous ground-mass which could be described with equal accuracy as a sandy limestone or as a calcareous sandstone. For complete accounts of the various groups geological textbooks must be consulted. We are concerned here only with their function as repositories of fossils.

Diagenesis

It must be appreciated that the geologist regards as a rock any material, hard or soft, which forms part of the Earth's crust. To him a clay or a sand is a soft rock, granite a hard rock. After material has been deposited it may suffer diagenetic change before it is converted or lithified into a hard rock. Water percolating through the pores of a sand may deposit part or all of its dissolved salts to fill the pore spaces and bind the sand into a sandstone. The prefix calcareous, ferruginous, etc., states the nature of the cement binding the quartz grains of a sandstone. The pore spaces of a clay rock are far too small to allow easy movement of water and therefore cementation is much less effective in argillaceous than in coarser-grained rocks. But when mud is deposited it may contain as much as 80 per cent of interstitial water. This interstitial water will be gradually squeezed out as additional rock material accumulates on top of the original bed. The mud is consolidated into a clay rock, with a considerable reduction in thickness. Any contained fossils will be flattened along the bedding planes. Peat is converted into coal by being buried beneath thousands of feet of strata. It is believed that a coal seam two feet in thickness was formed from the compression of a layer of peat thirty feet or more in thickness. The grains of a sand, on the other hand, are already in contact, so little further compression is possible.

Water percolating through a bed may dissolve material, particularly calcareous material such as shells, so that a sand which originally contained much shell debris may gradually have this removed until finally the sand is quite unfossiliferous.

Salts dissolved from one part of a bed may be deposited in some other layer. The layers of flint nodules in the Chalk are believed to have been formed soon after its deposition, when the pH of the waters circulating through the Chalk was fluctuating between that which dissolved the disseminated silica and that which caused the silica to be deposited to form a layer of flint. In clays concretions, usually partly calcareous, are formed, often round organic remains, which are thus protected from further compression.

Diagenetic changes may be selective. The matrix of a limestone composed of a fine-grained mosaic of calcite crystals may be silicified or dolomitized (dolomite is $MgCo_3$.

$CaCO_3$), while calcite shells are unaffected. The 'rottenstones' of Derbyshire were originally shelly limestones. After selective silicification the calcitic fossils were dissolved away to produce a hard siliceous rock with numerous cavities showing the former position of shells. With complete silicification or dolomitization, all trace of the original nature of a rock will disappear.

On stagnant parts of the sea floor the supply of available oxygen is exhausted by decaying organic matter. As there can be no replenishment of oxygen, the breakdown of organic matter continues by slow destructive distillation with loss of the volatile constituents and the final production of uncombined or 'fixed' carbon. *Anaerobic* bacteria flourish in such a reducing environment. Through the action of these bacteria, sulphides, particularly iron sulphide, FeS_2, are formed. The iron sulphide may be precipitated as a finely divided black pigment throughout the sediment, it may be concentrated as nodules of marcasite or pyrite, or it may replace the shells of organisms to produce the pyritized fossils which are to be found in the London Clay and in many other clays.

The coarse *Rudaceous Deposits*—originally beach shingle, river gravels, or the fan-deposits fanning at the foot of mountains in semi-arid regions—were laid down under such tumultuous conditions that any dead organisms other than the stronger bones of vertebrates would quickly be disintegrated. Cemented rudaceous deposits are referred to as conglomerates if the bulk of the pebbles and boulders are well rounded, and as breccias if the majority are markedly angular. Breccias can also accumulate in caves. Cave-breccias may contain the bones of animals that inhabited or were washed into caves. Examples are the cave deposits of Sterkfontein in South Africa, which have yielded the remains of the *hominoid* australopithecines, and the 'hyena den' of Kent's Cavern, Torquay.

Arenaceous Deposits may have formed in a wide range of environments: as dunes, either desert or coastal, as sandbanks in rivers and estuaries, or as the result of the in-filling of channels on delta plains, around the edges of lakes or more extensively on the continental shelf. They are dominantly composed of quartz grains within the size range of 2 mm to 0·1 mm in diameter and may have a porosity as high as 47 per cent. If cemented they are called sandstones. Arenaceous deposits are frequently unfossiliferous. This may be due either to the absence of life in the environment of deposition or to the solution effects of water which can easily percolate through them.

Argillaceous Deposits, made up not of quartz but of complex aluminium silicates, the Clay Minerals, may also have been deposited in a wide variety of environments. They may have been laid down as muds in the less agitated parts of lakes, on the continental shelves and the upper parts of the continental slopes, in the quieter reaches of rivers or in lagoons. Such environments are usually rich in life, and the extremely fine-grained nature of the clay and its imperviousness favour preservation. As the contained water is squeezed out, mud compacts into clay and this either into shale, if the bedding planes are closely spaced, or into mudstone, which is less well-bedded.

Calcareous Deposits, containing more than 50 per cent of calcium carbonate or of dolomite, are an extremely varied group of rocks. They may originate as organic accumulations, either shell banks or reefs formed in shallow water, or as oozes which mantle the lower parts of the continental slopes. They may also have been deposited partly or wholly as chemical precipitates, for the solubility of calcium carbonate, unlike

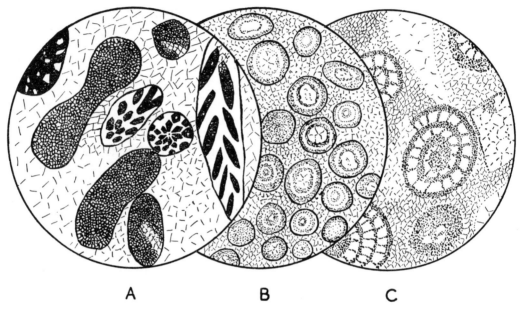

FIG. 9. Thin-sections of limestones

A Bioclastic limestone composed mainly of crinoid ossicles and bryozoan debris, which have been heavily impregnated with haematite. Bryozoan Bed, Rhiwbina, Cardiff; B Oolitic limestone, Bath; C Nummulitic limestone, Ghizech, Egypt (See p. 62)

that of most other salts, decreases with rise of temperature, so that either calcite or aragonite, the two forms of calcium carbonate, can be precipitated from saturated warm water. Limestone deposits can also form in caves. The textural range of limestone is great (Fig. 9). *Biohermal* limestones are reef deposits, composed mainly of organic remains; bioclastic limestones are made of shell debris set in a fine-grained calcareous matrix; oolitic limestones are built up of spherical ooliths, formed when bottom currents—agitated organically or chemically—precipitated calcareous muds, causing shell fragments, sand grains and other nuclei to roll backwards and forwards and become coated with concentric layers of calcareous material. Chalk is a variety of extremely fine-grained limestone composed mainly of coccolith debris, whilst the equally fine-grained oozes of the ocean depths are composed of the shells of foraminifera.

Evaporite Deposits are unfossiliferous, for they have been formed as precipitates from supersaturated brines.

Ferriferous Rocks are sedimentary rocks with an unusually high iron content. They were formed in specialized environments, either shallow-water marine or freshwater, but these were not usually inimical to life.

Siliceous Deposits are dense cryptocrystalline rocks which may contain silicified fossils. The siliceous earths, diatomite and radiolarite, are composed of the minute skeletons of diatoms and radiolaria mixed with some clastic mineral.

Phosphatic Deposits are of marine origin. They occur either as seams of phosphatized nodules in a sandy matrix or as beds of phosphatized limestone. A nonsequence or an unconformity usually underlies the nodule beds. These were formed during a period of very slow or inhibited deposition when shells, the teeth and bones of fish, etc., lying unburied on the sea floor, were replaced by or enclosed in chemically precipitated phosphate.

Of the *Carbonaceous Rocks*, the Coals were formed in a swampy environment, favourable to the growth of prolific vegetation. The peat produced was later compacted by the weight of overlying rocks and converted, by the gradual loss of its more volatile constituents, first into Brown Coal (Lignite) and then into the series of Bituminous Coals and finally, perhaps, to Anthracite. Bacterial action during the initial stages may have had a profound effect on the type of coal finally produced. As described on p. 12, many seams of *Humic* Coal are underlain by fossil soils showing the coal is found *in situ* or in the position of growth. The Cannel Coals, on the other hand, were formed from drifted plant remains. Cannel Coals do not occur as seams of very great extent; they are not underlain by *seat earths* and often contain elements of an upland flora not present in the Humic Coals. The Cannel Coals grade into the Boghead Coals and the Oil Shales. Under the microscope these are seen to contain great numbers of translucent yellow bodies, which are the remains of oil-secreting algae.

We must include the rocks yielding natural or 'mineral' oil, which is a complex mixture of hydrocarbons of the paraffins, naphthenes and aromatic series. Mineral oil is volatile and very liable to migrate into 'reservoir' rocks, usually sands or sandstone of high porosity or well-fissured limestones. For oil to accumulate, the reservoir rocks must be overlain by an impermeable cap rock, otherwise the migrating oil will reach the surface of the ground either to form a seepage or to evaporate, leaving behind deposits of viscous asphalt or bitumen. Mineral oil is of organic origin; it formed in marine, and probably also freshwater, evironments, when sediments with a high organic content were laid down under anaerobic conditions. The changes by which the organic matter was converted into the petroliferous hydrocarbons are not fully understood, but it is believed that the activities of bacteria played an important, if not essential, part.

Finally, the *Pyroclasts*: these are formed when debris, blown out of volcanoes, falls either on land or in water. Near the volcanoes coarse agglomerates, with boulders often several feet in diameter, will be deposited. Agglomerates sometimes contain fossiliferous rocks from the sides of the neck of the volcano. For example, blocks of fossiliferous chalk have been found in the material thrown out from a volcano on the island of Arran. No chalk outcrops on the island, but there can be no question that the Chalk underlies part of the island. Fine-grained volcanic ash falling into a sea or a lake may carry down with it and preserve organic remains. Exquisitely preserved butterflies and other insects have been found in beds of volcanic ash which fell into lakes at Oeningen in south Germany, whilst better known are the mummified bodies found during the excavations of the ash which buried the cities of Pompeii and Herculaneum when Vesuvius erupted in A.D. 79.

II

The Stratigraphical Table and the Absolute Dating of Rocks

IN THE early years of the last century the Stratigraphical Table was hammered out on the rocks of western Europe. By applying the Law of Superposition, the pioneer geologists arranged the beds in their order of deposition. They then divided the great column of rocks into groups, each characterized by a distinctive assemblage of fossils. Just as the historian thinks of the last thousand years or so in terms of dynasties and reigns, so the geologist divides the vastly greater period of time with which he is concerned into eras and periods. The rocks laid down during a period are known as a system.

The geologist recognizes five Eras. Working back from the present—the Quaternary Era which has only just begun—we speak of the Tertiary, Secondary and Primary Eras or, using another terminology, based on the fossil content of the beds, we refer to the Caenozoic (recent life), Mesozoic (middle life) and Palaeozoic (ancient life) Eras. It is customary today to use the terms Tertiary, Mesozoic and Palaeozoic, the alternatives having fallen into disuse, though Caenozoic is still a common term in America.

The nomenclature of the systems is of diverse origin. The Quaternary Era is usually divided into the Recent or Holocene (wholly recent) Period and the older Pleistocene (most recent) Period. The systems of the Tertiary—Pliocene (more recent), Miocene (less recent), Oligocene (little recent) and Eocene (dawn of the recent)—were named after the percentage of recent shells found in the beds. Most American and continental geologists regard the lower Eocene rocks as forming a separate system, the Palaeocene (ancient recent).

The Mesozoic systems were not christened so logically. The Cretaceous was named after the chalk (Latin *creta*), its most characteristic rock-type; the Jurassic, after the Jura Mountains, the site of much of the pioneer work on these rocks; and the Triassic after the three-fold division of its strata in Germany.

For the Palaeozoic rocks the terms Permian (after Perm on the River Volga in Russia), Devonian and Cambrian (Cambria, the mediaeval name for Wales) were obviously chosen from the places where the existence of these systems was first recognized. The Silurian and Ordovician Systems were named after Celtic tribes who lived in the type-area of the Welsh Borderlands and fought valiantly against the Romans. The term Carboniferous is clearly derived from the coal which is so widely found in its rocks. American geologists would separate the Carboniferous into two systems, the Pennsylvanian above and the Mississippian below, but this usage is not favoured in Europe.

The subdivision of systems into smaller units, stages and zones will be discussed in Chapter V.

The early geologists grouped together all the rocks beneath the Cambrian System[1] as the Pre-Cambrian Series. At that time fossils were known only from the Cambrian and later rocks. As knowledge of the Pre-Cambrian rocks increased it was recognized that they are greatly thicker, and must represent a much greater period of time, than all the later systems combined. Further, traces of fossils, though sometimes very debatable, were discovered in Pre-Cambrian strata. In many parts of the world it is possible to recognize an upper group of unmetamorphosed Pre-Cambrian rocks resting on strongly metamorphosed beds. The terms Algonkian or Proterozoic (earlier life) and Archaean or Archaeozoic (primaeval life) are used for these two groups.

For over 150 years geologists have realized that geological time must be of immense duration, but until very recently there had been no reliable method of determining the absolute age of rocks, that is, in millions of years. Estimates were made, based on the thickness of the sedimentary rocks divided by an assumed rate of sedimentation, or on the salt content of the oceans, which was believed to have been derived from the erosion of the lands, or on the rate of cooling of the Earth from its original molten state. The wide variation of the ages deduced, ranging from a few score to hundreds of millions of years, showed how many unknowns there were in the majority of the calculations.

But with the discovery of radioactivity the position changed. It was found that each place in the Periodic Table of the Elements is filled not by one element but by a number of isotopes having the same chemical properties but with different atomic weights. As well as the stable isotopes, which had long been known, a large number of radiogenic isotopes were discovered. For example, there are four isotopes of lead with atomic weights 204, 206, 207 and 208 respectively and of these only lead 204, 'common lead', is not generated by radioactive processes.

The absolute dating of rocks or minerals is dependent on finding specimens containing one or more of the radioactive isotopes uranium^{235}, uranium^{238}, thorium^{232}, rubidium^{87} or potassium^{40}. Such radioactive minerals are not stable like ordinary minerals but undergo spontaneous disintegration, owing to the emission of electrically charged α particles. At each emission new radioactive substances are formed and the process

---

[1] The term Pre-Phanerozoic is being used increasingly in place of Pre-Cambrian. Phanerozoic (clear evidence of life) is applied to the beds of Cambrian and later age, in which fossils occur abundantly.

continues along a chain of changes until a stable end product is produced. The radioactive isotopes of lead are the end product of the disintegration of thorium and uranium minerals, strontium^{87} of rubidium^{87} and argon^{40} of potassium^{40}, so one speaks of thorium-lead, rubidium-strontium, potassium-argon, etc., methods. The period of disintegration of each of the radioactive substances is known. It varies at different parts of the chain from a fraction of a second to millions of years. The disintegration proceeds steadily, a definite number of atoms being affected in each unit of time and not in a series of sudden complete transformations from one form to the next. The proportion of the end product present in a radioactive mineral is therefore a measure of how far it has proceeded along the chain to complete alteration, and hence of the time that has elapsed since that particular mineral was formed. The technological difficulties of analysis are very considerable, but they have now been largely overcome, so that radioactive ages, in millions of years, can be given with a high degree of confidence.

But the difficulties are not simply those of analysis. Not only are radioactive minerals extremely rare, but absolutely fresh material must be used for age determination. Many of the earlier age determinations have had to be discarded, as the material used was not fresh enough. Again, the radioactive age has to be tied into the Stratigraphical Table, and that is frequently the most serious trouble of all. The potassium-argon method is the only one that can be used for sediments. The other methods require minerals which occur solely in igneous rocks or in the mineral veins associated with igneous activity. It is all too often impossible to state the geological age of igneous rocks with precision. The Dartmoor Granite of south-west England cuts across and metamorphoses rocks of early Upper Carboniferous age. Unfossiliferous rocks, believed to be of fairly early Permian age, rest unconformably on the folded Carboniferous rocks. The age determination of micas from the Dartmoor Granite by both the rubidium-strontium and the potassium-argon methods agrees very well at 290,000,000 years, and this figure can be placed in the Stratigraphical Table within the narrow limits given above. But elsewhere one may find radioactive minerals occurring in veins cutting perhaps Carboniferous rocks, but without any younger beds present to fix an upper limit.

With the discovery of the potassium-argon method it was hoped that fossiliferous sediments containing potassium-rich minerals such as glauconite could be dated with precision on both scales, but unfortunately so far this method has not proved completely reliable. Sufficient fixed points have, however, been found for the Stratigraphical Table to be dated as shown on page 28.

A considerable number of determinations are available from Pre-Cambrian rocks; certain Archaean rocks from the extreme north-west of Scotland are older than 2,000,000,000 years. This figure has been substantially exceeded in the Archaean rocks of Canada, Rhodesia and the Kola Peninsula, northern Russia. The age of the Earth itself is believed to be approximately 4,500,000,000 years.

Another method, Carbon^{14}, is available for dealing with Holocene and late Pleistocene times. The isotope Carbon^{14}, one of the products of cosmic radiation, is associated in very small amounts with Carbon^{12} in carbon dioxide. Owing to the rapid circulation of the atmosphere, this radioactive carbon is incorporated in the atmospheric carbon reservoir. It is absorbed by living organisms, but on the death of an organism its supply of radioactive carbon is cut off. The Carbon^{14} in the organism, then, behaves like the radioactive minerals mentioned above, but unfortunately its rate of disintegration is extremely rapid, so that Carbon^{14} dating cannot be carried further back than 70,000 years and, even at a lesser age, the technical difficulties of measuring such minute amounts of radioactivity are very great. Carbon^{14} dating is therefore of great value to the archaeologist and others concerned with the last 10,000 years of the Holocene period, but is of strictly limited value to the geologist.

Stratigraphical Table (after Geological Society of London, 1964)

Era	Period	Beginning of Period in millions of years
Quaternary	Pleistocene	$1 \cdot 5 \pm 2$
Tertiary	Pliocene	7
	Miocene	25
	Oligocene	40
	Eocene	70 ± 2
Mesozoic	Cretaceous	135 ± 5
	Jurassic	180 ± 5
	Triassic	225 ± 5
Palaeozoic	Permian	270 ± 5
	Carboniferous	350 ± 10
	Devonian	400 ± 10
	Silurian	440 ± 10
	Ordovician	500 ± 15
	Cambrian	600 ± 20

III

The Nature and Preservation of Fossils

THE term fossil (Latin *fossilis*: 'dug up') dates from the sixteenth century. It was originally applied to any curious object, whether a crystal, prehistoric implement or fossil in the modern sense, that was found buried. But since then the term has been restricted to cover only the remains, or traces, of the life of the past. The fragments of pottery, metal tools or wooden furniture dug up by the archaeologist when he is excavating a site of a few thousand years' antiquity are not, therefore, regarded as fossils. They are too recent, for they date from a time when climatic conditions were little different from those of today. So our definition of fossils can be amplified to 'remains, or traces, of organisms that inhabited the Earth before the close of the last Ice Age', that is, roughly 10,000 years ago.

When an organism dies, whether on land or in water, its remains are exposed to the agents of organic decay and to the attack of scavengers. It is only under the most exceptional conditions, when it is sealed up before or immediately after death, that an organism can be preserved relatively unchanged.

The pebbles of amber that occur in the beaches of the south coast of the Baltic have long aroused human interest. Finds of them in Neolithic and later graves have been one of the clues in reconstructing the old trade routes between the Baltic and the Black Sea and so to the rich eastern Mediterranean lands. Today amber pebbles are prominently displayed in the gift shops of Copenhagen. Amber is fossilized resin, extruded in Tertiary times from coniferous trees. Insects were occasionally trapped in and covered by the resin; today they can be seen preserved in the transparent honey-coloured amber. So perfect is the preservation that one specimen of a spider showed traces of its thread.

Before World War I there was a flourishing trade in fossil ivory from Siberia. The ivory was obtained from the tusks of the mammoth (*Mammuthus primigenius*) which inhabited the steppe and tundra lands to the south of the ice-sheets towards the close of the last Glaciation. In a few cases complete carcasses of mammoths have been found. The famous Berezovka mammoth, now preserved in the Leningrad Museum, must have been feeding along a river bluff which collapsed beneath his weight. The shock of the fall broke his forelimb and forced a great clot of blood into his chest. His body was quickly frozen in the soft deposits of the flood plain and since then it has been preserved in the permafrost. The contents of his stomach suggested from the type of plants present that his death occurred in the autumn.

But such finds are most exceptional. With the vast majority of fossils, only the hard parts (the shells, bones, etc.) have been preserved. Such fossils can be easily picked out of uncompacted rocks, but if the 'host rock' has been well lithified, then the shells, etc., may be difficult to extract. The matrix obscuring parts of the fossil will have to be removed, according to the degree of hardness, with a brush, needle, mechanical mallet or dental drill. Acids that will attack only the matrix can also be used.

But fossils may undergo further change whilst embedded in the rocks. The early Pleistocene shelly sands, the 'Crags', exposed in the rapidly receding cliffs of East Anglia, contain a rich molluscan fauna. At first glance these shells, though somewhat iron-stained, appear as perfect as recent shells washed up on the beach below. On closer inspection, many of the shells are found to be friable and show signs of flaking on their outer surface. Such shells are composed of aragonite, the less stable of the two forms of calcium carbonate. The calcite shells of oysters, etc., in the same bed are not so affected. In beds of greater age the rarity or absence of organisms which secreted aragonite shells is probably an extension of this process.

Petrifications

Fossils, whether calcareous or carbonaceous, may be altered by the effects of water percolating through the rocks. In the simplest case, pore spaces originally filled by organic matter in life may be infilled by precipitated mineral matter. Thin-sections of the 'Bryozoan Bed' of the basal Carboniferous Limestone of the Cardiff neighbourhood show this very well, for red haematite has been deposited in the pore spaces of the calcite *bryozoan* and crinoid debris (Fig. 9). If the pores of a calcite shell are infilled by secondary calcite, then its original structure may be lost.

A more profound change occurs when a fossil is petrified. In petrification complete replacement takes place, the original hard parts being replaced, perhaps by silica which may, very exceptionally, be in the form of precious opal; perhaps by compounds of iron, including iron pyrites; or by phosphate, to name the commoner petrifying minerals. Petrifactions can be the most exquisite of fossils, for the detailed structure is now preserved in a material which is both more durable and more easy to handle than the original calcite, aragonite, carbon or *chitin*. A biohermal limestone of Permian age from Texas has yielded, after suitable treatment with acid, a magnificent assemblage of silicified fossils, including extremely spinose brachiopods with all their long spines in place. If these shells had been preserved in calcite in a calcite matrix it would have been impossible to extract them in their entirety.

On Big Horn Mountain, in Yellowstone National Park, U.S.A., are preserved the remains of twenty-seven forests which had been successively overwhelmed by pyroclastic material thrown out from nearby volcanoes. Many of the trees were standing upright. Silica, from percolating water, has replaced the stumps and logs, which are now more resistant than the ash beds and so stand out on the mountainside. Wood, cones, etc., petrified in this way often have their original cellular structure most beautifully preserved, as can be seen when the hard 'stone' is cut and polished. 'Coal balls' are another type of petrification, found in the coal-bearing beds of Upper Carboniferous age. In this case masses of plant debris have been impregnated with calcium and magnesium carbonate, or with iron sulphide, and now form hard concretionary masses in coal seams. Again preservation is often extremely good, as is seen in the *cellulose peels* that can be made from polished sections of coal balls.

At Herne Bay and in the cliffs of the Isle of Sheppey fragments of pyritized wood and cones occur in the London Clay, as well as marine shells such as those of Nautilus. Pyritized ammonites are not infrequent in the Cretaceous and Jurassic marine clays, whilst pyritized graptolites, relatively uncrushed, can be found in some of the Lower Palaeozoic shales. These pyritized fossils, often beautifully iridescent, are relatively heavy and, as at Herne Bay and Lyme Regis, may be washed out of the cliffs and swept together by the waves in heaps on the beach. But a word of warning—iron pyrites, particularly if it has been in contact with sea water, is liable to be oxidized to powdery sulphate after a few years' exposure to the air. Such pyritized fossils must be protected by dipping them in molten wax.

Petrifactions are not always as perfect as, perhaps, has been suggested above, particularly when fossils have been phosphatized. The fossil may form the nucleus around which further phosphate has been deposited to produce a shapeless mass of black phosphate with perhaps a recognizable fragment of a fossil in one corner. Prolonged search is usually necessary to find good specimens in phosphatic nodule beds.

Moulds and casts

Percolating water, instead of replacing organic material, may dissolve it away. If the walls of the cavity thus produced are strong enough, then a mould of the original fossil is left. By infilling the mould with Latex rubber, or one of the compositions used by dentists, a cast or replica of the original organism, can be made. To facilitate extraction of the cast from the mould it is essential to use a material which not only makes a sharp cast but is also flexible. The main building stone in the quarries on the Isle of Portland is overlain by a few feet of a very distinctive limestone, the 'Roach', full of the moulds of fossils. The external moulds show the ornamentation of the shells of the lamellibranchs (mainly *Trigonia*, Fig. 31 C) and of the turreted gastropods, whilst the internal moulds show to perfection the dentition of the trigonias. Most people are familiar with the casts in flint of sea urchins which can be found on beaches formed from the destruction of chalk cliffs, or in river gravels composed mainly of flint pebbles, or sometimes lying on the surface of the Chalk Downs. If such specimens are not too worn, lines of little raised dots can be seen on the upper surface. These mark the position of the pores of the tube feet in the *ambulacral* areas and show that one is looking at a natural cast. The calcite shell of these urchins was only a millimetre or two in thickness. After the death of the sea urchin its *test* became completely infilled with silica. The thin layer of calcite was later dissolved away to leave a perfect natural cast.

This kind of material can be of great help to the palaeontologist, for the details of the internal side of the shell are revealed. If the inside of a lamellibranch is infilled with hard matrix, perhaps harder than the fossil, then removing the matrix to reveal the internal features of the shell can be a most exasperating and difficult task. In the case of vertebrates the discovery of natural casts may be still more valuable. Our knowledge of certain early reptiles and amphibians which lived in Permian and Triassic times, at Lossiemouth and Elgin in north-east Scotland, is based largely on moulds of their bones in sandstones. By infilling these moulds with Plaster of Paris the shape of the bones can be reconstructed. Still more useful are natural casts of skulls, for the inside of the skull reveals the shape of the outer membrane of the brain and the position of the major blood vessels and nerve roots. The discovery of such material in the early Devonian rocks of Spitzbergen afforded a unique glimpse of the cranial features of the primitive ostracoderms (p. 48). The impressions of the brain, nerves and blood vessels were not only very similar to those of the modern lamprey, but, in addition, a number of massive trunks were present, interpreted as nerves supplying probable electric organs.

Films and impressions

Fossils may also occur as carbonized films on bedding planes. The volatile components of plants, especially of leaves, and of animals such as graptolites and arthropods with a skeleton of chitin plus protein, have gradually disappeared until only a film of carbon is left. The preservation of the detail of the leaves may be extremely good, but in the case of invertebrates it is very difficult to reconstruct satisfactorily material that has been squashed flat. As more and more carbon is lost, the films pass into impressions, but with a really fine-grained rock, laid down under very quiet conditions, these may be amazingly sharp.

The well-known Lithographic Stone of Upper Jurassic age, quarried at Solenhofen in Bavaria, has yielded impressions of jelly-fish, whilst an even more remarkable bed of shale high up on the side of Mount Stephen in the Canadian Rockies has given us a glimpse of the soft-bodied creatures that lived in the Middle Cambrian seas over 500,000,000 years ago. This shale, the Burgess Shale, contained impressions of medusoids, holothurians and annelids, as well as several arthropods of unknown affinities.

Trace fossils

Finally, the traces of former life. On the bedding planes of red sandstone of Triassic age in Cheshire and elsewhere are sometimes to be found the footprints of tetrapods. Vertebrate footprints are usually fairly easily recognizable, provided they are well enough preserved, but more difficult to interpret are the wide variety of tracks, trails and burrows that occur on the bedding planes of Palaeozoic and later rocks. Some of these have been proved to be the trails of trilobites. The term trace-fossil is used to include a wide variety of markings and structures found in sedimentary rocks; these may well be due to the activity of animals moving on or through sediment during its deposition, but to reconstruct the soft-bodied animals which made the traces is extremely difficult and debatable. Coprolites are the fossilized excrement of vertebrates, whilst the name gastroliths is given to polished pebbles, which may well have served as the gizzard stones of the large Mesozoic reptiles. In some cases gastroliths have been found placed very suggestively amongst the disarranged bones of a reptilian skeleton.

Micropalaeontology and palynology

During the last thirty years vastly increased attention has been paid to the study of microfossils, that is of fossils so small that they must be studied under the microscope. This great interest in micropalaeontology has come largely from the oil industry.

The majority of oil borings today are rotary borings. The hole is drilled by rotating a bit, armed with industrial diamonds, at the end of an ever-increasing length of steel tubing. Mud of controlled viscosity passes down the drill stem, lubricates the bit and then returns to the surface in the space between the drill stem and the sides of the hole (Fig. 10). The mud carries upwards fragments of the rock cut loose by the bit. These fragments or cuttings can be collected by passing the mud through a baffle box or by using a power-driven shaker. Complete samples of the rock which is being drilled through can be obtained by withdrawing the drill stem to the surface and replacing the normal bit by a core barrel, a hollow steel cylinder armed at its lower end with cutting wheels. This is lowered down the hole, drilling recommences and slowly the core barrel is filled with a cylinder of rock, which can then be lifted to the surface for examination. Coring is clearly much slower and therefore more expensive than ordinary drilling. The amount of coring to be carried out in any one hole must be seriously considered, for the additional scientific data obtainable has to be weighed against cost and time.

Cores are only a few inches in diameter, so the chances of finding megafossils in them are relatively slight. But cuttings, as well as cores, may contain microfossils, that is foraminifera, ostracods, conodonts, together with embryonic megafossils and disarticulated fragments of sponges, brachiopods, echinoids, fish, etc. (Fig. 11). The microfossils can be isolated from the matrix by suitable treatment. Soft clay rocks may be disintegrated by soaking them in water. Shales or mudstones require boiling in sodium bi-carbonate or sodium hydroxide, or may have to be heated and then plunged into a powerful detergent. Arenaceous, siliceous or chitinous microfossils can be isolated from calcareous rocks by the use of suitable acids. A dried residue of mineral grains, rock fragments and microfossils is finally produced, and out of this the microfossils are handpicked and mounted for microscopic examination. The same techniques can be used for the breakdown of specimens collected from surface outcrops. Using

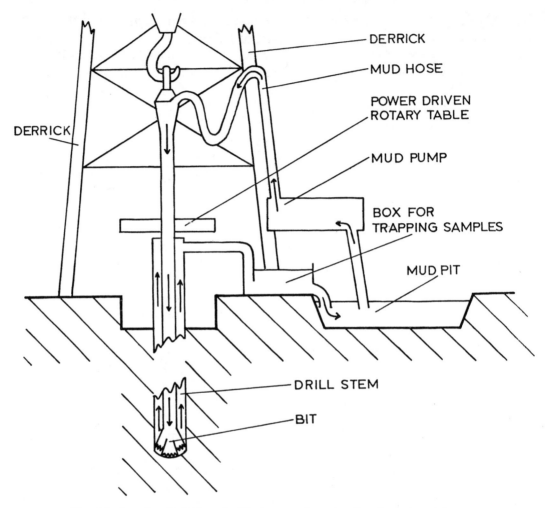

Fig. 10. An oil-well drilling rig. The arrows show the direction of mud flow

a sufficiently high-power microscope it is possible to study such minute organisms as fossil dinoflagellates, coccolithophorids, etc., as well as extremely minute forms, a few microns in diameter, whose biological affinities are uncertain.

Rocks may also contain the spores and pollen of plants. These can be isolated by techniques much more complex than those given above, and examined by a palynologist, a botanist or palaeobotanist who has specialized in the study of spores and pollen. As will be shown in Chapter VII, *palynology* is of great value in the study of Quaternary and Pleistocene deposits, as well as of the coal-bearing strata. Increasing use is being made of palynology in the field of applied geology, for it gives the micropalaeontologist another weapon in his investigations of the environments of the past and of the relationships of the beds in which oil is being sought.

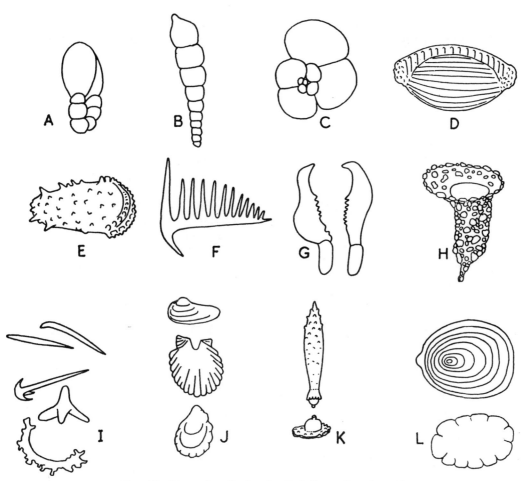

FIG. 11. Examples of microfossils (all greatly enlarged)
A Foraminiferan (*Endothyra*); B Foraminiferan (*Alveolina*); C Foraminiferan (*Dentalina*); D Foraminiferan (*Globigerinoides*); E Ostracod; F Conodont; G Scalecodont; H Radiolarian; I Sponge spicules; J Young lamellibranchs; K Echinoid spines; L Otolith and fish scale

The incompleteness of the palaeontologist's material

The material which the palaeontologist has at his disposal is often not too promising. In the first place, only very exceptionally is there any indication of the form and nature of the soft parts of the organisms. Examples of such freak preservation are specimens of nuculid lamellibranchs from the Lias of Gloucestershire containing moulds of the coiled intestines; of a trochid gastropod from the Lower Albian greensands of Kent showing a mould of the gut; of an annelid from the Tremadocian Shales of Shropshire with the form of the gut, the jaws and details of the skin recognizable; and of a dibranchiate cephalopod from the Upper Cretaceous of Syria with indications of the stomach, intestines and other organs. Restorations of vertebrates may be morphologically correct, but the form, colour and indeed nature of the outer covering is only too often merely an intelligent guess. But for the frozen specimens of mammoth we might never have known for certain that these elephants were covered with long hair, though this might have been inferred from other evidence of their climatic environment. Only a few localities have yielded impressions of the skins of the great dinosaurs, whilst again there are few known impressions of the stream-lined body of the marine swimming reptiles, the ichthyosaurs (Fig. 34).

The majority of fossils are forms which are now extinct. If closely similar types are living, then we can use them for placing the extinct organisms in their correct systematic

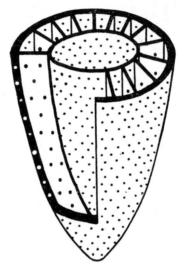

Fig. 12. An archaeocyathid
(after Okulitch)

position as well as deducing their mode of life. These problems become extremely difficult when one has to deal with long-extinct forms without any living relatives. For example, in the Cambrian rocks of many parts of the world are thick limestones made up of the skeletons of archaeocyathids. These organisms did not build true reefs, but seem to have formed extensive 'gardens' in relatively narrow belts parallel to the coastlines of shallow seas. Figure 12 shows the structure of a typical archaeocyathid with its porous inner and outer walls connected by porous radially disposed partitions. No archaeocyathids survived beyond the close of the Cambrian Period. Their systematic position is naturally uncertain. They have been grouped by different workers with corals, sponges, protozoans and calcareous algae. Cogent arguments have been advanced against each of the claims. In the latest review it is proposed that the archaeocyathids form an independent phylum and in the absence of any knowledge of their soft parts this seems the most reasonable course to take.

Many fossils have suffered either distortion or compression. Distortion of fossils is one of the results of tectonic stress. When fine-grained rocks are laterally compressed they develop a cleavage, that is, new planes of parting, which may be at a high angle to the original bedding. The bedding may be indicated by colour banding or by layers of slightly different texture. Rocks with a well-developed cleavage will not split at all easily along the bedding planes. It is only too easy to break across the fossils instead of along the planes on which they lie, so the hammer must be skilfully wielded if fossils are to be found in cleaved rocks. But even if the cleavage is parallel to or nearly parallel to the bedding, the fossils may be very considerably elongated in one direction (Fig. 13).

The skull of a vertebrate animal compressed by the weight of the overlying strata, and perhaps originally lying half on its side, will need highly skilled interpretation. Comparable difficulties occur with invertebrates, particularly those with weak skeletons, composed of chitin and *tanned proteins*. As an example we can cite the graptolites,

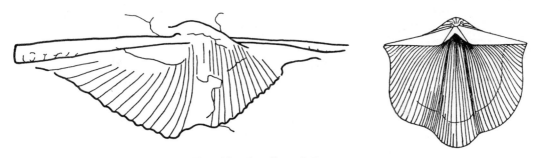

FIG. 13. The effect of cleavage

On the right is an undeformed specimen of the brachiopod, *Spirifer verneuli*. On the left is a distorted specimen, a so-called 'Delabole butterfly', from the Devonian beds of the Delabole slate quarries, Cornwall

which are a group of fossils of great value to geologists in interpreting the Lower Palaeozoic rocks. Graptolites normally occur as carbonized films on the bedding planes of shales. They were colonial forms, the colony consisting of one or more branches (stipes) carrying either unilaterally or bilaterally arranged rows of cups or thecae which housed individual zooids. In such flattened forms the stipes often appear toothed, like miniature fretsaw blades (Fig. 17). When only material of this kind was available for study the graptolites were usually grouped with the Coelenterata. Shortly before World War II, R. Kozlowski found fragmentary graptolites preserved uncrushed in chert nodules in the Tremadocian rocks of central Poland. The chert was dissolved away with hydrofluoric acid, the fragments mounted and sectioned with a microtome. This magnificent material showed that in these early dendroid graptolites there were three separate types of thecae arranged in alternating triads along a *stolon*: the periderm had an inner layer with growth bands and growth lines, and an outer layer of laminated tissue (Fig. 14). Kozlowski claimed that these features strongly suggested that the affinities of the graptolites are not with the Coelenterata but with the pterobranchs of the phylum Chordata. Most palaeontologists have accepted his views, but many zoologists do not regard the case as proved. These important discoveries were ready for publication when war broke out. Although Professor Kozlowski's department was destroyed in the fighting in Warsaw, his specimens, text and illustrations were fortunately preserved, and in 1948

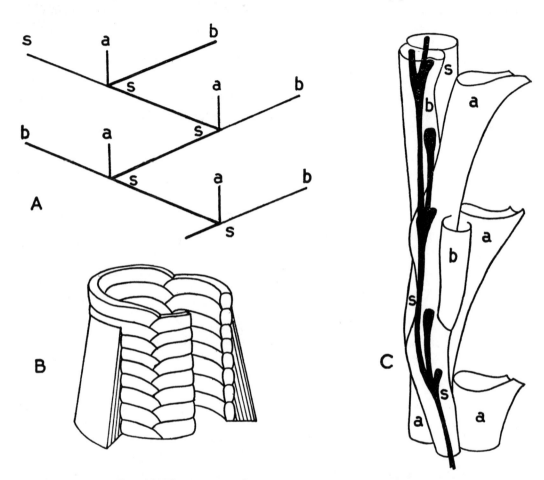

FIG. 14. The structure of dendroid graptolites (after Bulman)
A The three kinds of thecae arranged in alternating triads; B Detail of chitinous exoskeleton showing alternating half rings of fusellan tissue surrounded by laminated cortical tissue; C Detail of stipe of *Dendrograptus* (stolon system in black) (X40); a—Autotheca, b—Bitheca, s—Stolotheca

his work was published. Pyritized graptolites preserved in the solid can sometimes be found, but these cannot be cut with a microtome and have to be studied by *serial grinding*.

A further difficulty is that much fossil material is dismembered and incomplete. This is not usually the case with the stronger invertebrates, corals, brachiopods, lamellibranchs, but can be a very serious hazard when dealing with arthropods, vertebrates and especially plants. A rather extreme example of this was the discovery by L. J. Wills of fragments of scorpions in sandy shales in the Triassic rocks of Bromsgrove, Worcestershire. The shale fragments were thoroughly dried and then placed in hot water, when they crumbled, so that the chitinous fragments could be removed with a needle or a camel-hair brush and mounted for examination. Reconstruction of the scorpions required most painstaking work, for whilst a complete scorpion has 120 sclerites, it was the exception in the Bromsgrove material to find more than one isolated sclerite at a time and at best four or five connected abdominal ones. Wills suggested that this high degree of dismemberment might be the result of cannibalism.

G. G. Simpson, the expert on fossil horses, stated in 1952 that whilst literally tens of thousands of single teeth, jaws, skull fragments and separate bones of *Hyracotherium* or *Eohippus* (the dawn horse) have been found in the Palaeocene strata of the western United States, only four skeletons, and none of them really complete, were preserved in museums.

The position is even worse for the palaeobotanist. Plant material is rarely found complete. All too often he is dealing with stems, leaves and fructifications all found separated and bearing different names; one problem is to attempt to relate them to one another. The result may be misleading, as is shown by a recent account of the cycads. Today the cycads, a small family of the gymnospermous seed plants, live in the Tropics. Great numbers of leaves, numerous stems and some reproductive organs very like those of modern cycads have been found in rocks of Mesozoic age; so many that the Mesozoic was named by some the 'Age of the Cycads'. With further work it was proved that certain of the plants called cycads and with cycadean-like stems and leaves had very different reproductive organs. So for these a new class, the Bennettitales, was established. It is now believed that more than half of the so-called Mesozoic cycads belong to the Bennettitales and that the number of confirmed cycads is very small.

The difficulties that confront the palaeontologist have been emphasized, perhaps over-emphasized, for at other horizons he may find well preserved material, particularly in marine strata, giving him a representative sample of a fossil population. He needs only one such sample to be able to reconstruct the life of that particular period of the past. There are still large areas of the continents which have not been explored systematically by palaeontologists. Techniques for obtaining core-samples of the deposits which have accumulated on the ocean floors during Pleistocene and earlier times have only recently been developed. It is certain that in the future many of the present gaps in the Palaeontological record will be filled by discoveries similar to and as exciting as those mentioned above.

IV

The Succession of Life

BRITISH geologists are singularly fortunate, for there is probably no other area in the world of similar size to the British Isles in which such a range of geological systems can be studied. But some systems, e.g. the Miocene, are unrepresented, for during this period the British Isles formed an upland area and if any Miocene beds were laid down they have been removed by subsequent erosion. And the rocks of some other systems, e.g. the Trias, are almost barren of fossils, for Britain was then a desert or semi-desert; in some parts of the world fossiliferous marine rocks of Triassic age are to be found. Elsewhere Triassic strata, though non-marine, have yielded quite good faunas, particularly of vertebrates.

Therefore, whilst in this chapter we shall be concerned primarily with British rocks, it will be necessary to fill in the gaps in our record by reference to some extra-British occurrences.

The Pre-Cambrian rocks

Except in the north-west of Scotland, where most of the rocks are strongly metamorphosed, outcrops of strata of Pre-Cambrian age are limited in Britain to a few small areas, such as Charnwood Forest in Leicestershire, the Malvern Hills, parts of Anglesey, the Wrekin and the Longmynd in Shropshire. Elsewhere Pre-Cambrian rocks are hidden beneath the mantle of younger strata. But in Canada, around the shores of the Baltic, South Africa, Siberia, Brazil, southern India, there are great 'Shield' areas, composed of Pre-Cambrian rocks.

The occurrence of fossils has been claimed at a number of localities, especially in the unmetamorphosed, or but slightly metamorphosed, Algonkian or younger Pre-Cambrian beds. Many of these claims have been hotly disputed. In certain areas graphite, a form of carbon, is quite abundant. By some it is regarded as of organic origin, by others as inorganic. In certain of the Pre-Cambrian rocks of Australia, Siberia and parts of the United States are beds of limestone showing a distinctive laminated and concentric structure (Fig. 15 A). These stromatolites, which occur as dome-like masses up to fifty-five feet in diameter and twenty-two feet in height, have been interpreted as formed by colonies of lime-secreting blue-green algae. Some dispute this and claim they are concretionary structures of inorganic origin. The stromatolites appear to resemble very

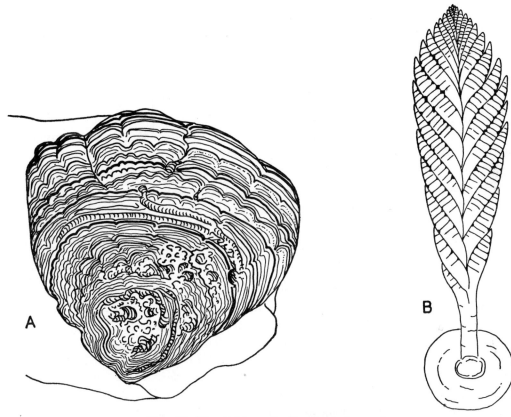

FIG. 15. 'Fossils' from the Pre-Cambrian rocks
A Stromatolitic nodule, Belt Mountains, United States, X½; B Reconstruction (after Ford) of frond-like impressions (*Charnia*) and disc-like impressions (*Charniodiscus*) found at Charnwood Forest, Leicestershire, X½

closely bodies of unquestioned algal origin found in the Palaeozoic rocks. In 1954 uncompressed spores and filaments of blue-green algae and fungi were detected in transparent chert from south Ontario. The presence of amino acids in this chert has been proved by organic trace analysis, whilst the age of the bed has been determined by the potassium-argon method as 1,700,000,000 years. In 1957 a schoolboy noticed frond-like impressions on a fine-grained cleaved tuffaceous siltstone of Pre-Cambrian age in Charnwood Forest, Leicestershire. These were associated with, but not joined to, disc-like markings. The likely explanation seems to be that these impressions were formed by a seaweed-like alga (Fig. 15 B), which was torn off the rocks, washed into a nearby pool and quickly buried, possibly by a shower of volcanic dust. Similar impressions found in south Australia have, however, been referred to the Octocorallia.

Though our knowledge of Pre-Cambrian life is still so incomplete and debatable, it is clear from the study of the Algonkian sediments, at least, that they were laid down in the same range of environments as the present and within the existing range of temperatures, that is from freezing point to approximately 100°F. Conditions favourable for organic life must therefore have persisted on the surface of the Earth for at least the past 2,000,000,000 years.

The Cambrian beds

Wherever in Britain the junction of Cambrian beds on Pre-Cambrian rocks is exposed it is an unconformity. The basal Cambrian beds are coarse beach sands and shingle (now cemented into hard quartzite) passing upwards into finer-grained marine (glauconitic) sandstone and then into thicker clays and silts, so we must picture an advance of the seas (a marine transgression) submerging a somewhat irregular surface cut across the Pre-Cambrian strata. The coarse basal beds are unfossiliferous, but fossils occur in the overlying sandstones and thin limestones. The majority of the fossils of the Lower Cambrian are chitinous; inarticulate brachiopods and a distinctive group of trilobites, the olenellids (Fig. 16), with elongated crescentic eyes set near the centre of the head shield, no *facial suture*, numerous and often spinose segments, and in most genera a long spine, resembling that of the modern *Limulus*. The olenellids must have had a long evolutionary history, but their ancestors are unknown. Presumably they were soft-bodied and therefore have not, as yet, been found. The burrows of worms are plentiful in some localities, notably in the 'pipe-rock' of north-west Scotland.

In the Middle and Upper Cambrian rocks the olenellids are replaced by other, much more varied, families of trilobites, and the first articulate brachiopods appear. It is worth noting that the inarticulate brachiopod *Lingulella davisii* (Fig. 16), preserved in great abundance on the bedding planes of hard shales in some Welsh localities, is almost identical with *Lingula*, still living in the muds off Japan. Reference has already been made to the unique soft-bodied fauna of the Burgess Shale (p. 32) and to the archaeocyathid limestones (p. 36).

The Ordovician System

On palaeontological grounds the Tremadocian Series (see Appendix) should be included with the Ordovician System, for its trilobite fauna is characterized by the appearance of a number of families which are typical of the Ordovician beds. The first graptoloids, the many-branched dendroids, are also found in the Tremadocian. In the overlying Arenigian Series the true graptolites developed from the dendroids. Then followed a rapid reduction in the number of branches, for the Llanvirian rocks are characterized by the two-branched 'tuning-forks'. Biserial graptolites (Fig. 17) are also present, but they become more important at higher levels. The Ordovician rocks of England and Wales are either shales and mudstones with graptolites or arenaceous rocks yielding a

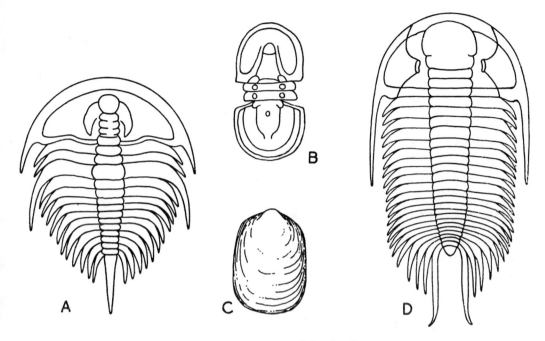

FIG. 16. Some Cambrian fossils
A *Olenellus*, a Lower Cambrian trilobite, X$\frac{1}{2}$; B *Agnostus*, a blind trilobite, Middle Cambrian, X7; C *Lingulella*, an inarticulate brachiopod, Upper Cambrian, X2; D *Paradoxides*, a spinose trilobite, Middle Cambrian, X$\frac{1}{4}$

rich variety of trilobites and articulate brachiopods. Limestones are very poorly developed. In extreme north-west Scotland, however, the Durness Limestone of lowest Ordovician age has yielded a rich molluscan fauna of gastropods and straight-shelled nautiloids, some very large. The Ordovician limestones of Estonia are famous for their fauna of echinoderms. The fixed *pelmatozoans* are represented by many-plated cystoids (Fig. 18), and also by crinoids with a many-armed cup borne on a long stalk, whilst the ancestors of the echinoids are also present.

The Silurian System
The general geographical setting of Silurian times was the same as in the Ordovician Period—a subsiding trough, a geosyncline, extending north-eastwards from central and northern Wales across the Lake District to the Southern Uplands of Scotland. Shelf sea deposits on the southern margin of the geosyncline are well developed in the Welsh Borderlands. In Ordovician times there was much volcanic activity in the central parts of the geosyncline, but this died out early in the Silurian Period.

The sediments of the geosynclinal areas are shales and mudstones with interbedded sandstones and grits, often contorted and ploughing into the underlying shales. These contorted sediments are interpreted as the result of turbidity currents plunging down the submarine slopes of the geosyncline and carrying with them coarse sediment which was

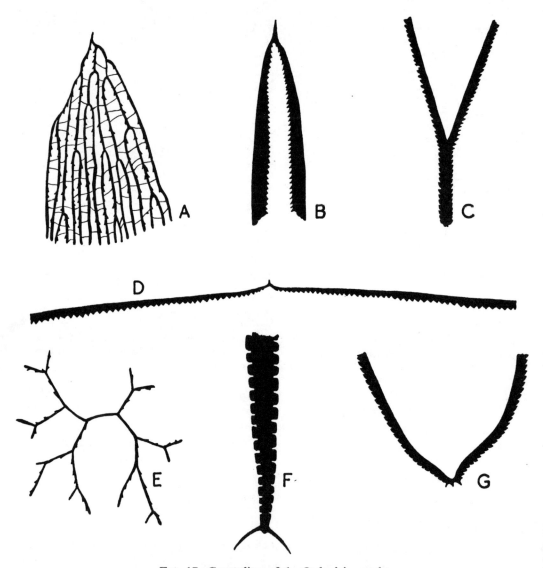

FIG. 17. Graptolites of the Ordovician rocks

A *Dictyonema*, dendroid graptolite, Tremadocian, X2; B *Didymograptus murchisoni*, 'tuning fork' graptolite, Llanvirnian, X2; C *Dicranograptus*, biserial stripe becoming uniserial, Caradocian, X2; D *Didymograptus extensus*, horizontal two-branched graptolite, Llanvirnian, X2; E *Dichograptus*, a many-branched graptolite, Arenigian, X2; F *Climacograptus*, biserial form with simple thecae, Caradocian, X2; G. *Dicellograptus*, flexuous reclined uniserial stipes, Ashgillian, X2

spread across the quietly deposited muds. The shales yield a rich variety of graptolites: biserial forms in the basal beds, but above these the uniserial monograptids showing great variation in the type of thecae (Fig. 19). The true graptolites did not survive the

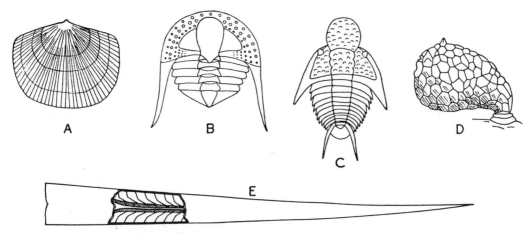

FIG. 18. Fossils of the shelly facies of the Ordovician

A *Orthis*, an articulate brachiopod, X1; B *Cryptolithus*, a bottom-living trilobite, blind and with broad flat head shield, X2; C *Staurocephalus*, a nektonic trilobite with a very inflated glabella, X$\frac{3}{2}$; D *Aristocystis*, a cystoid, X$\frac{1}{3}$; E *Endoceras*, a straight nautiloid, up to 6 ft. in length. Note the large siphunicle and backward-pointing septa, where the outer shell is shown as worn through

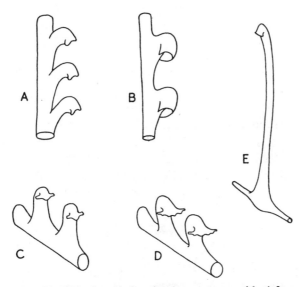

FIG. 19. Thecal variation in the monograptids (after Bulman)

A *Monograptus priodon*, Wenlockian; B *M. lobiferus*, Valentian; C *M. triangulatus*, Valentian; D *M. spiralis*, Valentian; E *Rastrites*, Valentian (all greatly enlarged)

Silurian Period, though the dendroids lingered on with little change into the Carboniferous. As in the higher Ordovician beds of the graptolitic *facies*, distinctive trilobites with very expanded eyes and inflated glabellas also occur in the shales and are interpreted as nektonic forms.

The fauna of the shelf seas is well known. In the Middle Silurian of Wenlock Edge in Shropshire, at the 'Wren's Nest' near Birmingham and elsewhere, a thin-bedded limestone occurs, the Wenlock Limestone. Slabs of this limestone form beautiful 'fossil graveyards', for they show, etched out by weathering, a great variety of bryozoans articulate brachiopods, trilobites, crinoid fragments, corals, etc. Interrupting the even bedding of the limestones and shales are irregularly shaped masses of solid limestone, 'ballstones' (Fig. 20). The ballstones, built up of masses of rugose and

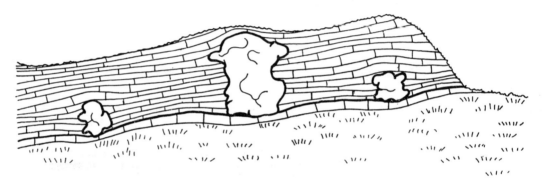

FIG. 20. Ballstones in the Wenlock Limestone

Sketch of quarry face showing three ballstones. The thin-bedded limestones and shales arch up over and are depressed beneath each ballstone

tabulate corals bound together with laminated sheets of stromatoporoids (an extinct group of hydrozoans) and much bryozoan material, are interpreted as small 'patch reefs'. The chief differences from the shelly faunas of the Ordovician are the greater variety of brachiopods, including the appearance of the spiriferids and terebratulids with calcified internal skeletons and the much increased importance of crinoids and corals. Silurian trilobites are well known—for example the rolled-up specimens of the 'Dudley Bug', *Calymene blumenbachi* (Fig. 21 A)—but the trilobites had passed their acme, for new families did not appear to take the place of those which had died out. In the higher beds of the Ludlovian this rich and varied marine fauna begins to disappear. Bedding planes may be found covered with fossils, but there is little variety, just a few species of brachiopods together with occasional molluscs. Conditions were changing from truly marine to brackish water.

The Caledonian Orogeny

Towards the close of the Lower Palaeozoic a major *orogenic* episode occurred when the rocks of the Lower Palaeozoic geosyncline were folded and intruded by granitic masses, such as those of the Lake District. The effect of this orogeny was to change completely the geographical setting of the British Isles. A range of mountains striking south-west to north-east developed across Scotland and the Lake District, and possibly North Wales, whilst central and South Wales and the Welsh Borderlands formed lower but still hilly country. Farther to the south a new geosyncline came into existence, stretching west to east from Brittany to the Rhineland. This chain of events de-

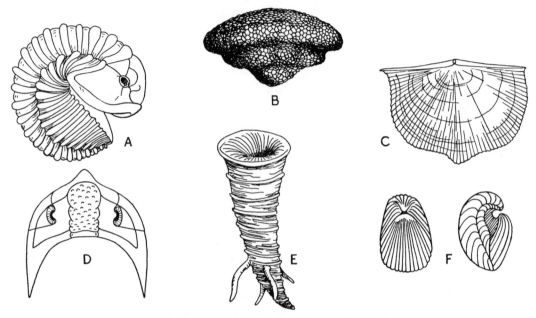

FIG. 21. Fossils of the shelly facies of the Silurian

A Rolled-up specimen of the trilobite *Calymene blumenbachi*, X$\frac{1}{2}$; B *Favosites*, a compound tabulate coral, X1; C *Leptaena*, a strophomenid brachiopod, X1; D Head shield of the trilobite *Dalmanites caudatus*, X1 (note the large eyes); E *Omphyma*, a solitary rugose coral, X$\frac{2}{3}$; F *Conchidium knightii*, a protrematous brachiopod, X$\frac{1}{2}$; the specimen on the right has split down the well developed median septum

termined the sequence and range of environments represented by the rocks of the Devonian and Carboniferous Systems.

The Devonian Period

The shelf seas of Devonian times did not extend north of the line of the present Bristol Channel until towards the close of the Period. Farther north the rocks of Devonian age are non-marine and of a very different facies—the Old Red Sandstone.

The marine Devonian rocks of south-west England were strongly deformed at the close of the Carboniferous Period by the Armorican Orogeny, so the rocks of the original type area yield but few fossils and these are often distorted (Fig. 13). Fortunately the marine Devonian strata of the Ardennes and the Rhineland are much more fossiliferous, so these areas provide the standard European succession, as is shown by the stage terms listed in the Appendix. The fauna of the marine beds is broadly similar to that of the Silurian rocks. Rhynchonellid and spiriferid brachiopods increase in importance, corals are more varied, trilobites continue to decline. An important change is the rise of the goniatites, coiled cephalopods with sharply angular suture lines.

The Old Red Sandstone was deposited in Scotland in inter-montaine basins and in Wales on delta plains on the margin of the shelf sea. Invertebrates, the inarticulate brachiopod, *Lingula*, and a few lamellibranchs, gastropods and ostracods are restricted

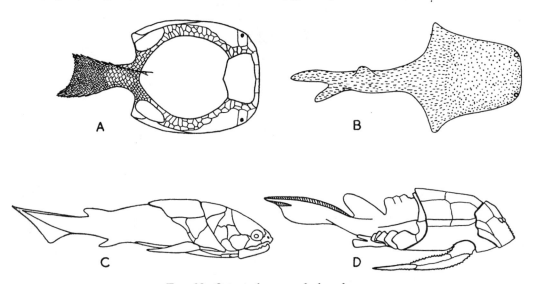

FIG. 22. Ostracoderms and placoderms

A *Drepanaspis*, a dorso-ventrally compressed ostracoderm from the Upper Old Red Sandstone, 1 ft. in length; B *Thelodus*, an early ostracoderm, 6 inches in length, without body armour, but with a continuous covering of minute scales. Highest Silurian and basal Old Red Sandstone. The Ludlow Bone Bed of Shropshire is largely made up of *Thelodus* scales; C *Coccosteus*, a Middle Old Red Sandstone placoderm, up to 16 inches in length; D *Bothriolepis*, a highly specialized placoderm from the Upper Old Red Sandstone, 1 ft. in length

to the basal beds; afterwards the water became too brackish. The Old Red Sandstone, is, however, famous for being the first formation to contain in some abundance both vertebrates and land plants. The lower beds yield agnathids, the ostracoderms, distantly related to the modern lamprey. The ostracoderms were jawless, without paired fins. Some forms had heavy body armour, others were unarmoured, but with numerous small scales. Fish-like creatures with jaws are also to be found. These placoderms, which did not survive the Palaeozoic, are a varied and primitive group, including both shark-like and heavily armoured bottom-living forms with jointed anterior appendages (Fig. 22).

At higher levels in the Old Red Sandstone occur Chondrichthyes (sharks) with a cartilaginous skeleton and Osteichthyes (bony fish) (Fig. 23) with a much more efficient jaw apparatus than the placoderms. The Osteichthyes are represented by ray-finned fish (Actinopterygii), lung fish (Dipnoi) and the lobe-fins (Crossopterygii).

At a number of famous localities, including Rhynie in Aberdeenshire and Kiltorcan in southern Ireland, *vascular* plants have been found. The Rhynie flora of Middle Devonian age, beautifully preserved in chert, consists of psilophytes, without true roots, the leaves being either small or absent. The Upper Old Red Sandstone of Kiltorcan and elsewhere has yielded a flora of spore-bearing lycopods with true roots and numerous leaves, and fern-like pteridophytes (Fig. 24).

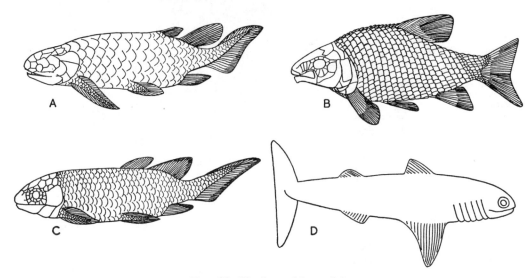

FIG. 23. Sharks and bony fish

A *Holoptychius*, Upper Old Red Sandstone, a crossopterygian 2½ feet in length; B *Lepidotus*, an actinopterygian 1 foot in length. Note the large thick scales; C *Dipterus*, an early dipnoan, Old Red Sandstone, 1¾ feet in length; D *Cladoselache*, an Upper Old Red Sandstone elasmobranch, up to 4 feet in length

Whilst vertebrates and vascular plants first appear in some abundance in rocks of Devonian age, evidence as to their pre-Devonian history is slowly accumulating. Fragments of ostracoderms have been found in sandstones of Ordovician age in Colorado, whilst recently digestion in acid of pieces of Wenlock and other British Silurian limestones has yielded ostracoderm scales and spines. Traces of vascular plants, some admittedly debatable, have been claimed from rocks as old as the reputedly spore-bearing Lower Cambrian clays of Estonia and Latvia.

The Carboniferous Period

A northward transgression spread across the now worn-down remnants of the Caledonian mountain chains. At times the sea reached as far north as central Scotland. In South and North Wales, in the Bristol area and in Derbyshire, the Lower Carboniferous is made up very largely of typical Carboniferous Limestone. A massively bedded pale-coloured limestone, in large part bioclastic, laid down in clear water, it yields a varied fauna of brachiopods, tabulate and rugose corals, crinoids, bryozoans, etc. (Fig. 25). Large solitary rugose corals and the gigantid productid brachiopods, up to a foot or more in width, are characteristic of the upper beds of the limestone. Further

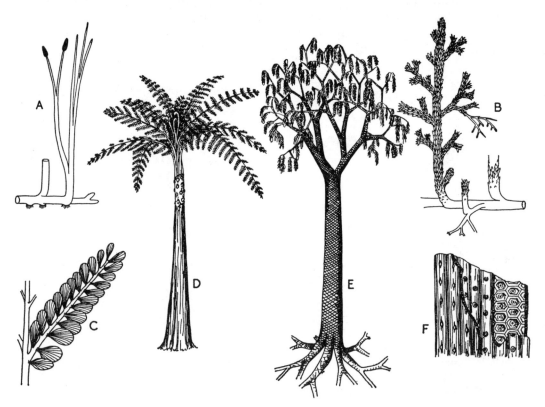

FIG. 24. Devonian and Carboniferous plants

A *Rhynia*, a psilophyte from the Middle Old Red Sandstone (after Kidston and Lang); B *Asteroxylon*, a psilophyte from the Middle Old Red Sandstone (after Kidston and Lang); C *Archaeopteris*, an Upper Devonian pteridophyte (after Schimper); D Reconstruction of an Upper Carboniferous tree fern about 25 feet in height (after Morgan); E Reconstruction of the Upper Carboniferous lycopod *Lepidodendron*; F Stem of *Sigillaria*, showing on the right the outer surface with leaf scars and to the left the effects of progressive decortication

north, in northern England and central Scotland, even the term Carboniferous Limestone Series is a misnomer. Thick clear-water limestones do occur, especially in the lower beds of the Pennines, but farther north the marine limestones are reduced to thin bands in a dominantly shaly succession with some sandstones. The shales are partly marine; at other *horizons* they are of brackish or freshwater origin, with a fauna of lamellibranchs, fish, ostracods and drifted plant debris. Coal seams of workable thickness underlain by seat earths are not uncommon. These indicate prolonged periods of plant growth in swamps on the surface of deltas spreading southwards from land areas to the north.

The Carboniferous Limestone Series is overlain by the Millstone Grit consisting of thick alternations of coarse deltaic grits containing drifted plant debris with, at intervals, coal seams too thin to be worth working, and of marine shales yielding a rich variety of goniatites and some thin-shelled pectinoid lamellibranchs (Fig. 26).

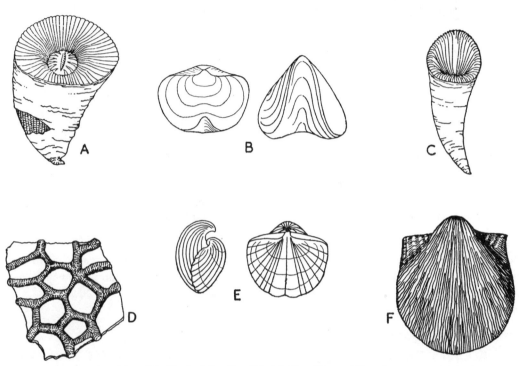

FIG. 25. Typical fossils of the Carboniferous Limestone

A *Dibunophyllum*, solitary rugose coral with complex axial column, X1; B *Pugnax*, rhynchonellid brachiopod, X1; C *Zaphrentis*, solitary rugose coral without an axial column, X1; D *Vaughania*, tabulate coral, X2; E *Spirifer*, telotrematous brachiopod, X1; F *Productus*, protrematous brachiopod, X1

In the overlying Coal Measures beds of marine origin are restricted to 'marine bands', often only a few feet in thickness, with productids, goniatites, crinoids, etc., overlying coal seams. The Coal Measures are a rhythmic succession, repeating time and time again the cyclothem shown in Fig. 27, with the freshwater shales making up the bulk of the beds. The coal seams are composed of many species and genera of lycopods, psilophytes and pteridosperms; essentially a swamp flora. Drifted fragments of conifers, which must have lived on higher ground, are also to be found.

The Coal Measure swamps were inhabited by insects, including dragonfly-like forms with a wing-span of two feet. Towards the close of the Devonian Period,

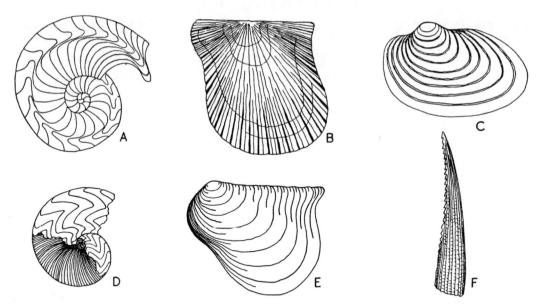

Fig. 26. Fossils of the Millstone Grit and the Coal Measures
A *Reticuloceras*, goniatite, Millstone Grit, X1; B *Pterinopecten*, lamellibranch, Millstone Grit, X2; C *Posidonia*, lamellibranch, Millstone Grit, X1; D *Goniatites*, Millstone Grit, X1; E *Naiadites*, lamellibranch, Coal Measures, X2; F Spine of *Ctenacanthus*, an elasmobranch, Coal Measures, X1

amphibians developed from the crossopterygian fish. The Coal Measures have yielded the remains of a variety of labyrinthodont amphibians, mainly clumsy tetrapods, as well as lepospondyls with snake-like bodies and very weak limbs.

The Armorican Orogeny

In late Carboniferous times southern England was affected by another orogeny. The Upper Palaeozoic rocks of Devon and Cornwall, South Wales and the Mendips were compressed from the south. The thick limestone bands were thrown into massive folds, with local thrusting, whilst the shales were cleaved and locally altered into slates. Into the folded rocks were intruded the granites of south-west England. In Midland and northern England and Scotland these earth movements produced little folding but much block faulting.

The New Red Sandstone

Once again the geography of Europe was profoundly altered as a result of a period of earth stress. The British Isles, as well as most of France, the Low Countries and Germany, formed a land area, whilst a new geosyncline, the Tethys, came into being approximately along the line of the present Mediterranean.

The continental deposits laid down in the British Isles during the Permian and Triassic Periods form the New Red Sandstone. They are usually barren of fossils,

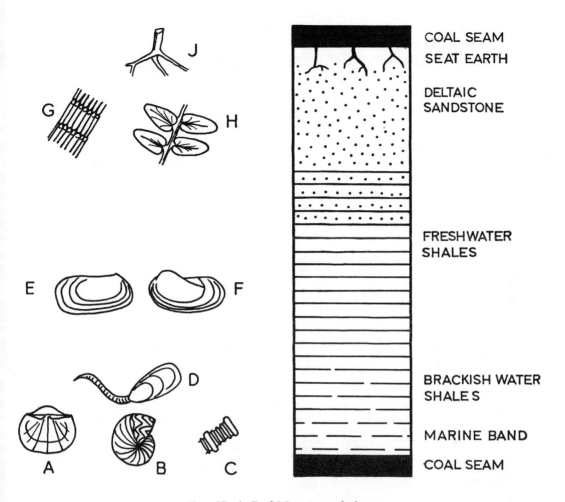

Fig. 27. A Coal Measure cyclothem

Typical fossils of the several units are: Marine Band: A *Productus*; B Goniatite; C Crinoid ossicles: Brackish water shales: D *Lingula*. Freshwater shales (lamellibranchs): E *Carbonicola*; F *Anthraconauta*. Deltaic Sandstone: G Stems of *Calamites*; H Leaves of *Neuropteris*. Seat Earth: J Stigmarian roots

though very occasionally, as at Bromsgrove (p. 39), fortunate finds have been made. The fossiliferous lenticles at Bromsgrove, yielding (as well as scorpions) the remains of lung-fish, labyrinthodonts, conchostracans and the stems and leaves of xerophytic plants, give us a glimpse of the life which went on in and around temporary ponds, similar to the modern 'vleys' of the semi-arid parts of South Africa. Only in north-east and north-west England and in Antrim are there marine deposits of Permian age. The Magnesian Limestone contains in its lower beds reef limestones, built up mainly by algae and fenestellid bryozoans, and yielding a limited fauna of productids, rhynchonellids and lamellibranchs. But the higher beds are unfossiliferous and the limestone passes upwards, and also laterally eastwards, into evaporite deposits. The Magnesian

Limestone was laid down near the margin of a sea which transgressed from the Tethys across eastern Russia, then round the northern edge of the Baltic Shield to Britain and Germany. This sea, the Zechstein Sea, soon became too saline to support life, and finally evaporated away. The salt deposits thus formed were covered by great sheets of sand—partly wind-blown, partly water-laid—the Bunter Sands, the lowest division of the German Trias. Another marine transgression, following a more direct route from the Tethys, then submerged much of Germany to deposit another rather abnormal limestone, the Muschelkalk. The fauna of the Muschelkalk differs markedly from that of the Zechstein Limestone, for it is not of Palaeozoic aspect but contains the forerunners of the rich and varied Jurassic life. The Muschelkalk is famous for its ceratitic ammonoids with the frilled lobes to their sutures (Fig. 28), for its forests of crinoids and for lamellibranchs related to *Trigonia*. The fauna is a specialized one and this limestone also passes upwards into evaporite deposits, which are overlain by continental deposits, mainly argillaceous, the Keuper Beds, the highest unit of the German Trias.

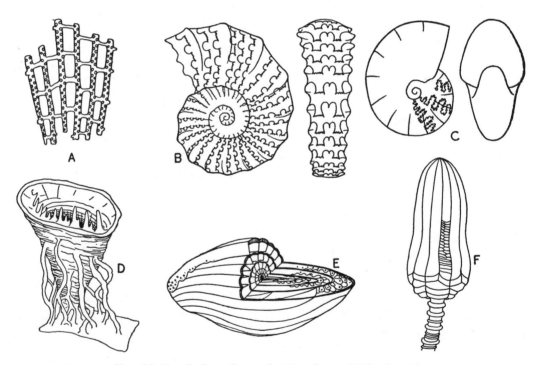

Fig. 28. Fossils from the marine Permian and Triassic rocks

A *Fenestella*, Permian, a bryozoan with a lace-like skeleton, X50; B *Ceratites*, Muschelkalk, showing the ceratitic suture line of frilled lobes and rounded saddles, X$\frac{1}{2}$; C *Ptychites*, Middle Trias, showing the ammonitic suture with frilling of both lobes and saddles, X$\frac{1}{8}$; D *Prorichtofenia*, Permian, an aberrant sessile brachiopod with the ventral valve of a coralloid shape whilst the dorsal valve (not shown) is lid-like, X1; E *Fusulina*, Permian: these 'giant' foraminifera range up to 2$\frac{1}{2}$ inches in length; F *Encrinus*, Muschelkalk, crinoid, X1

The Marine and Permian Trias

In Texas, in Russia to the west of the Urals and in the eastward continuation of the Tethys along the present line of the Himalayan chains, a normal (not a specialized) marine development of the Permian is preserved. These beds are characterized by distinctive foraminifera (the fusulinids) and ammonoids, together with many brachiopods, including the richtofenids, which had a coralloid habit, sponges, corals, etc. (Fig. 28). In the marine Permian rocks are to be found the last of the trilobites, of the blastoids, of many of the Palaeozoic orders of crinoids and of the rugose and tabulate corals.

The marine Triassic Beds, best known from the Alps and the Himalayas, yield a very different fauna. Hexacorals have taken the place of the rugose and tabulate corals. Only one family of the Permian ammonoids (with relatively simple suture lines) survived into the Trias and from this family developed a great variety of ammonites, including the ceratites, with extremely complicated suture lines and a great diversity of shell ornamentation. Ancestral belemnoids and cidaroids (an important group of regular echinoids) are also known from the Trias. Brachiopods declined markedly in importance, in sharp contrast with the rise of the lamellibranchs.

The Karroo System

The most fossiliferous successions of continental beds of Permian and Triassic age are found in South Africa. Resting unconformably on the Pre-Cambrian rocks of the African Shield are non-marine beds of great thickness. The Karroo System commences

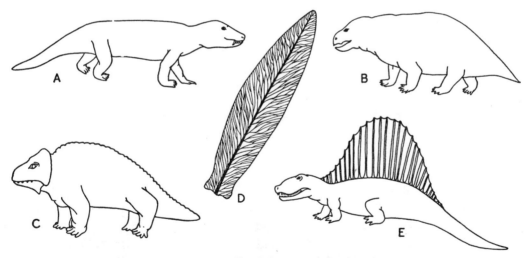

Fig. 29. Fossils of the Karroo System

A *Dicynodon*, an herbivorous synapsid: the skull is up to 3 feet in length; B *Cynognathus*, a carnivorous synapsid, 4–5 feet in length; C *Pareiasaurus*, large herbivorous cotylosaur, up to 10 feet in length; D Frond of *Glossopteris*, $X\frac{3}{4}$; E *Dimetrodon*, a large carnivorous pelycosaur, up to 10 feet in length. The function of the dorsal 'sail' is unknown. It may have been a temperature—regulating device

with beds of unquestionable glacial origin, the Dwyka Tillite, overlain by the Ecca Shales, yielding a distinctive assemblage of plants, the *Glossopteris* flora which is also found above or associated with glacial beds in India, South America, Antarctica and Australia.

The overlying Beaufort and Stormberg Series, clay rocks and sandstones, contain a

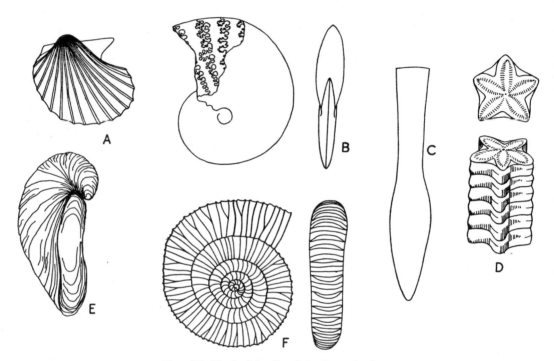

FIG. 30. Typical fossils of the Jurassic clays

A *Oxytoma*, a lamellibranch, Lias and Oxford Clays, X1; B *Oxynoticeras*, a smooth laterally compressed ammonite. The ammonitic suture is visible where the outer wall has been worn through, Lower Lias, X1; C *Belemnites*, a dibranchiate cephalopod, occurs in all clays, X1; D Stem ossicles of the crinoid, *Pentacrinus*, Lias and Oxford Clays, X2; E *Gryphea*, a sessile lamellibranch, Lower Lias, X½; F *Dactylioceras*, a loosely coiled ammonite, with prominent ribbing, Upper Lias, X½

magnificent assemblage of labyrinthodont amphibians, cotylosaurs (the reptiles with the closest affinities to the amphibians), synapsids (mammal-like reptiles) and thecodonts (the ancestors of the dinosaurs) (Fig. 29). No fewer than 600 species of reptiles have been found in the Beaufort Series. Both carnivorous and herbivorous forms are present, but there is very little trace of the vegetation on which the vast numbers of herbivores must have browsed. Locally, layers yield lung-fish and crustaceans. Some of these amphibians and reptiles occur in beds of proved Permian or Triassic age in Europe and America, enabling the faunas of the northern and the southern continents to be correlated.

The Jurassic Period

Towards the close of the Triassic Period the irregular land surface which resulted from the earth movements of late Carboniferous times must have been reduced to very low relief. Another transgression spread from the Tethys, submerging most of the British Isles, France and Germany. The Jurassic rocks of Britain consist essentially either of clays or of limestones, often oolitic. The clays contain a great variety of ammonites,

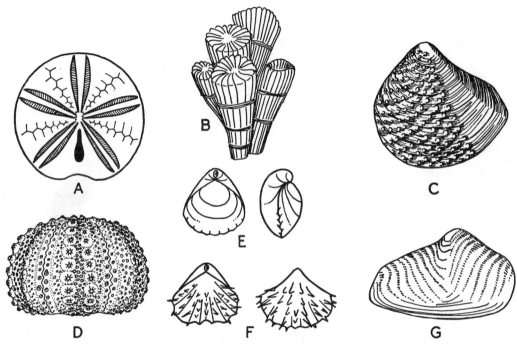

FIG. 31. Typical fossils of the Jurassic limestones

A *Clypeus*, an irregular echinoid, Inferior Oolite, X½; B *Thecosmilia*, a compound hexacoral, Corallian, X½; C *Trigonia*, an active lamellibranch with different ornamentation on anterior and posterior parts of the shell, Portlandian, X¾; D *Hemicidaris*, a regular echinoid with prominent bosses for spines. Corallian, X1; E *Terebratula*, a telotrematus brachiopod, Inferior Oolite, X1; F *Acanthothyris*, a spinose rhynchonellid brachiopod, Inferior Oolite, X3; G *Goniomya*, a burrowing lamellibranch, Cornbrash, X¾

belemnites, lamellibranchs, including many oysters, some crinoids, mainly pentacrinids (Fig. 30), and very occasionally the bones of the great swimming reptiles, the ichthyosaurs and the plesiosaurs, and even of the flying reptiles, the pterosaurs. In the limestones, on the other hand, ammonites are much rarer, lamellibranchs more varied, including species of *Trigonia* together with many burrowing forms. Rhynchonellid and terebratulid brachiopods are numerous and varied. Another important element of the limestone fauna is the abundance and variety of irregular echinoids, which had developed from *Cidarids* (Fig. 31). In Britain, reef-building hexacorals are restricted to a few horizons and to a few genera, but farther south, in the Jura Mountains, there are massive limestone reefs built up of a great variety of hexacorals and calcareous sponges. Towards the close of the middle Jurassic there was a regression of the sea, when deltaic deposits spread down from the northern land-mass into Yorkshire and beyond. These deltaic deposits have yielded many remains of gymnospermous plants, including cycads, bennettitales and ginkgos (Fig. 32). At the close of the Jurassic Period there was an

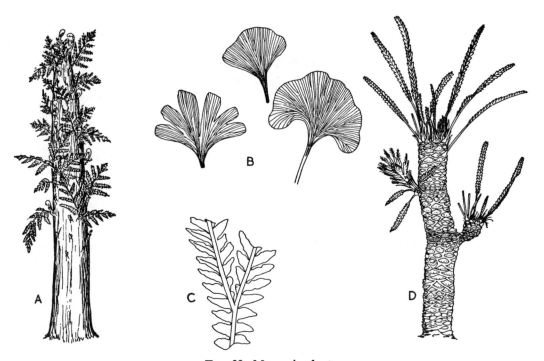

FIG. 32. Mesozoic plants

A Restoration of the tree-fern *Tempskya*, probably about 20 ft. high (after Andrews and Kern); B Leaves of species of *Ginkgo* (after Brown); C Leaf of *Thinnfeldia* (after Jones and de Jersey); D Reconstruction of *Williamsonia*, Bennettitales: trunk was probably less than 6 feet in height (after Sahni)

even more complete regression of the sea, when the non-marine Purbeck and the basal Cretaceous Wealden Beds were laid down. The Wealden Beds have yielded numerous dinosaurian remains. Mammalian remains, chiefly fragments of jaw and teeth, had been known from only a few horizons in the Jurassic, but within the last few years some fissures in the Carboniferous Limestone of the Bristol neighbourhood and Glamorgan have been discovered, infilled with brecciated material of earliest Jurassic age. These fissures have yielded a great wealth of teeth, skulls and long bones of primitive mammals. The Jurassic mammals were all small, unlike many of the contemporary reptiles, and as yet are imperfectly known. They include docodonts (eotherians), triconodonts of doubtful systematic position, multituberculates and pantotherians.

The Cretaceous Period

The Cretaceous deposits of the British Isles tell the story of a most extensive marine transgression. The non-marine Wealden Beds are overlain by variable marine beds, the Lower Greensand. These pass upwards into clays, the Gault, and that into a unique limestone, the Upper Cretaceous Chalk. This is the succession in southern England and it also records the gradual recession of the shores of the sea, until during the deposition of the Upper Chalk in Senonian times almost the whole of the British Isles must have been submerged. From other parts of the world there is similar evidence of an Upper Cretaceous transgression, the greatest extension of shelf seas across the continental masses in geological times.

The invertebrate fauna is characterized by the variety of ammonites, including uncoiled forms, and of belemnites. Both these groups became extinct at the close of the Period. Lamellibranchs continued to rise in importance. In the Tethyan regions the rudistids flourished, giant lamellibranchs in which one valve was greatly developed and almost coralloid in appearance, whilst the other was small and lid-like. Other typical groups are the irregular heart urchins or spantangoids (Fig. 33) and amongst the foraminifera, the globigerinoids (Fig. 11).

Another most important feature was the appearance of the angiosperms, the flowering plants, in the Lower Cretaceous. This was followed by an explosive radiation, so that by Upper Cretaceous times they had supplanted the gymnosperms as the dominant group of plants. Recent studies using the electron microscope have shown that the Chalk is composed dominantly of coccoliths and coccolith debris, so that these planktonic flagellates must have flourished in unusual abundance.

The Cretaceous was also the period when the reptiles reached their acme, both in size and specialization. The flying pterosaurs attained a wing-span of nearly thirty feet. Cretaceous plesiosaurs and ichthyosaurs were even more specialized for a marine life than their Jurassic predecessors. On land lived a great variety of dinosaurs, known best from the continental deposits of America and East Africa. Bipedal forms included the great carnivorous theropods and the herbivorous ornithopods, whilst amongst the quadrupedal dinosaurs were the gigantic swamp-dwelling sauropods, the stegosaurs with their tiny heads and bizarre plated backs, and the horned ceratopsians (Fig. 34). Yet none of these dominant reptiles survived the close of the Cretaceous Period. The exact cause of their extinction is unknown but must be related, in part at least, to the changes in geography caused by the great Cretaceous marine transgression and by the changes in food supply due to the rise of the angiosperms. But this would affect only the land dwellers. Disease has also been suggested, but so far little evidence of this has been obtained from fossil material.

The Tertiary history of America and Australia

In late Cretaceous times a phase of powerful earth movements, the Laramide Revolution, uplifted north-south striking mountain ranges along the western side of North America. To the east of, and in down-faulted basins between, the mountain ranges a great thickness of continental beds were deposited. These have yielded to collectors in the well exposed 'badlands' a magnificent series of mammalian remains. In the basal Tertiary Beds, the Palaeocene, can be traced the rise of the placental mammals. At first, generalized forms like the carnivorous creodonts and the herbivorous condylarths (Fig. 35), then in the Eocene appear the true ungulates, both artiodactyls and perissodactyls,

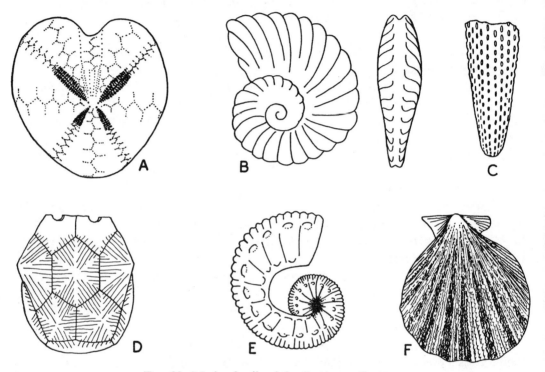

FIG. 33. Marine fossils of the Cretaceous System

A *Micraster*, irregular spatangoid echinoid. Note the petalloid ambulacral areas, Chalk, X1; B *Hoplites*, ammonite, with prominent ribs which do not cross the venter, Gault, X1; C *Ventriculites*, sponge, Chalk, X2; D *Marsupites*, free swimming crinoid, Chalk, X$\frac{2}{3}$; E *Discoscaphites*, an uncoiling ammonite, Chalk, X$\frac{3}{2}$; F *Aequipecten*, lamellibranch, pectiniform with strongly corrugated shell, Upper Greensand, X$\frac{3}{4}$

whilst the fauna of the Oligocene is of a more modern type due to the disappearance of many of the small-brained generalized archaic groups. In many groups of mammals the same sequence of changes can be traced: increase in size, increase in the capacity of the brain and gradual specialization of teeth and feet. In Miocene times there was a broad up-arching of the Rocky Mountains, the Cascadian Uplift, associated with much block faulting and igneous activity. This caused the change to a more arid type of climate which is reflected, for example, in the horse lineage by the gradual disappearance of the broad-footed, shallow-toothed 'browsers' and the rise of the faster-moving 'grazers' with their deep-crowned teeth. The Miocene fauna was modernized

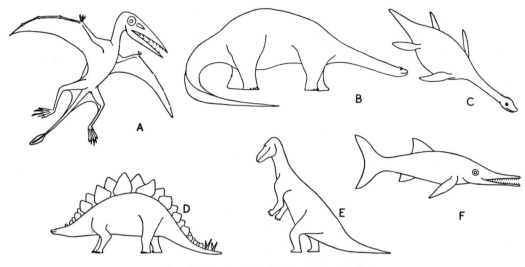

FIG. 34. Jurassic and Cretaceous reptiles

A *Pterosaurus*, a flying reptile. Jurassic forms like this had a wing-span of about 2 feet, but in some of the tailless Cretaceous forms the wing-span was as much as 25 feet; B *Diplodocus*, a swamp-dwelling sauropod, 80 feet in length; C *Plesiosaurus*, a marine swimming reptile, up to 40 feet in length; D *Stegosaurus*, an armoured dinosaur, 25–30 feet in length; E *Iguanodon*, a bipedal herbivorous dinosaur, up to 30 feet in length; F *Ichthyosaurus*, a marine swimming reptile, 10 feet or more in length

FIG. 35. Some Tertiary mammals

A *Phenacodus*, condylarth (primitive ungulate), Palaeocene, 5½ feet in length; B *Eobasileus*, amblypod (primitive herbivore), Eocene, about 8 feet in length; C *Dromocyon*, creodont (primitive carnivore), Eocene, 3–4 feet in length; D *Megatherium*, ground sloth, Pleistocene, up to 20 feet in length; E *Hyracotherium* (*Eohippus*), Palaeocene, the first of the 'horse' series, 1½ feet in length

by the disappearance of many unsuccessful groups which were unable to adapt themselves to life in a prairie environment. The North American Pliocene deposits, laid down under still more arid conditions, are but sparingly fossiliferous, but fortunately rich Pliocene vertebrate localities are known from Europe and Asia. The Pliocene fauna is essentially of modern type with 80 per cent of its families of terrestrial mammals still extant.

The Tertiary faunas of Australia and South America were, however, very different. In Australia, isolated from Asia, probably in the late Mesozoic, the marsupials were able to develop as carnivores and herbivores without competition from the placentals until very recent times. The communication between North and South America was broken early in Tertiary times. In South America a number of unique orders of ungulates developed, together with edentates and marsupials. When communication was reopened in Pliocene times the majority of the specialized South American faunas were not able to compete with the southward-migrating modern types of ungulates and carnivores, though in the Pleistocene certain South American forms, including the ground sloths and the armadillos, spread northwards across the isthmus into North America.

The Tertiary rocks of Europe

The general geography of Europe in Tertiary times was very different from that of the Mesozoic. The rocks laid down in the Tethyan geosyncline were folded in mid-Tertiary times and then uplifted to form the 'young mountain chains', the Pyrenees, the Alps, the Apennines and the Carpathians. The Tethys was separated from a new basin of deposition, the North Sea Basin, by a central European land area. Intermittently communications opened between the Tethys and the North Sea, through Aquitaine and then round Brittany and along the English Channel. The margins of the North Sea Basin fluctuated repeatedly and considerably, so the Eocene strata of south-east England, north-east France, the Low Countries and north-west Germany consist of wedges of marine sands and clays towards the present North Sea coasts, interfingering away from the coasts into non-marine sands and clays with local developments of freshwater limestone. In Oligocene times the sea spread across northern Germany and Russia, so the British Oligocene Beds are very largely of non-marine origin. In late Oligocene times came the main folding of the Alps. Farther northwards there was widespread uplift, so no beds of undoubted Miocene age are to be found in England, and indeed it was not until well into the Pliocene Period that deposition recommenced to form the basal 'Crags' of East Anglia.

The fauna of the marine Tertiary rocks of south-east England such as the London Clay is largely molluscan; gastropods and lamellibranchs, but no ammonites and belemnites, since they were extinct. At intervals beds yielding species of the 'giant' foraminiferan, *Nummulites*, occur (Fig. 36). These nummulitic horizons indicate periods of communication with the Tethys, for there the Tertiary Beds contain numerous foraminifera. In the Tethyan marine limestones, numerous sea urchins, corals and pectens also occur; these are rare in the deposits of the muddier waters of the North Sea Basin. Only in the Pliocene 'Coralline' Crag of East Anglia is a clear-water assemblage preserved, characterized by its abundance of bryozoans (which were grouped in the past with the corallines).

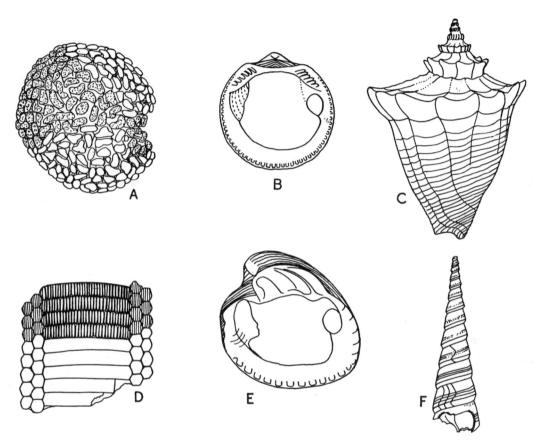

FIG. 36. Fossils of the English Tertiary Beds
A *Meandropora*, bryozoan, Coralline Crag, X½; B *Glycimeris*, lamellibranch with taxodont type of hinge, Lower London Tertiaries, X1; C *Athleta*, a highly ornamented gastropod, Barton Beds, X1; D Tooth-plate of the ray, *Myliobatis*, Barton Beds, X½; E *Venericor*, a lamellibranch with heterodont dentition, Bracklesham Beds, X½; F *Turritella*, a siphonostomatous gastropod, Bracklesham Beds, X1

The non-marine Tertiary Beds yield an invertebrate fauna that is predominantly molluscan. Many similar gastropods and lamellibranchs are living today, so it is possible to infer whether or not they indicate brackish or freshwater conditions. Remains of angiosperms are plentiful at some horizons, for example the 'Water Lily Bed' of the Oligocene of the Isle of Wight, whilst on the Continent extensive beds of peat occur, now in the form of Brown Coal (Lignite). To the north-west of Cologne the main lignite seam reaches a thickness of 100 metres. Vertebrate remains, especially in the older Tertiary rocks, are mainly in the form of disarticulated fragments swept into fissures and hollows on the contemporary land surface, very different from the highly fossiliferous bedded deposits of the United States.

The Pleistocene Period

We are not concerned here with the problem of the number and extent of the several Pleistocene glaciations. It is sufficient to recognize that there were a number of glacial episodes, when ice-sheets from the gathering grounds on the Scandinavian mountains spread, at times, as far south as the line of the Bristol Channel–Thames–lower Rhine–Harz Mountains. During the interglacial episodes the Scandinavian mountains may well have been bare, or as bare as they are today, of ice. To the south of the ice-sheets was tundra, passing into steppe and then into temperate forest. As the ice-sheets waxed and waned, these climatic belts swung backwards and forwards across Europe. But each advance of the ice ploughed off the unconsolidated deposits, including those of the preceding interglacial episode, over which it advanced. It is only in sheltered places that interglacial deposits are preserved, whilst our record of the older glacial episodes is mainly obtainable from those areas where they were not overrun by a later glaciation. We have, however, a fairly complete record of the deposits laid down by the last glaciation and of the melting back of the ice-sheets, during latest Pleistocene and Holocene times, to their present limits.

The vertebrate fauna of the Pleistocene differs from that of late Pliocene by the appearance of *Elephas*, *Bos* and *Equus* and also by the rare traces of *Homo* and his much more plentiful implements. The vertebrates can be grouped into tundra, steppe and *taiga* assemblages, whilst we can also recognize a broad chronological division in each *biotope* into Older, Middle and Late Pleistocene. The general effect of this period of such marked fluctuations of climate was one of impoverishment. The recent mammalian fauna is relatively restricted, owing to the extinction of so many Pleistocene 'giants' such as cave bear, Irish Elk, mammoth and giant sloth, whilst the range of many of the survivors was much reduced; for instance, the horses disappeared from America until they were reintroduced by man in the sixteenth century, and rhinoceroses and hippopotami used to inhabit much of Eurasia.

The invertebrates known from marine Pleistocene deposits round the continental margins and from beds laid down in lakes and rivers are very similar to the modern forms.

It is the same story with the plants, but these have great value as indicators of climatic conditions, especially in the recognition of interglacial deposits and of the retreat stages of the last glaciation.

V

The Use of Fossils as Time-Markers

IN CHAPTER II the Stratigraphical Table was briefly set out (p. 28) and in Chapter IV we traced, in outline, the main differences in the forms of life found in the rocks of the successive systems, as well as the broad geographical conditions under which each system was deposited. We have not, however, discussed the principles by which a geologist subdivides the great pile of sedimentary rocks or by which he traces time planes and correlates beds so that he can say that rocks of a particular lithology in one area were laid down during the same period of time as rocks of a different type in another area. If he can thus correlate the successions of different areas he is in a position to make palaeogeographic deductions. Further, if he can trace, through folded and faulted strata, distinctive beds, which were deposited as horizontal layers, then he has unravelled the effects of later earth movements. For many economic purposes, such as the search for oil, coal and ore-deposits, a detailed knowledge both of the structure of the beds and of their conditions of deposition is of the greatest importance. Therefore the use of fossils as time-markers is not only an academic exercise. The oil companies in particular employ hundreds of palaeontologists, especially micropalaeontologists, to study the fossils obtained from surface exposures and trial borings by the exploration geologists and from the wells that are sunk during the exploitation of oil-fields.

Before discussing critically the value and limitation of fossils as time-markers it will be necessary to review briefly the way in which the geologist's terminology of formations, stages and zones has developed.

Formations

Fossils have for long been collected by man. Specimens of 'shepherds' crowns', flint casts of the irregular sea urchins *Discoidea* and *Conulus*, have been found in Neolithic graves. Later, fossils were regarded by the Church with strong suspicion. They were 'sports of the Devil' placed in the rocks to mislead those who were unwilling to accept them as due to Noah's flood. That versatile genius Leonardo da Vinci (1452–1519) and some enlightened seventeenth- and eighteenth-century philosophers and scientists fully appreciated the nature of fossils as relics of life of the past; Guettard (1715–1786), for example, compared trilobites with crabs and lobsters. But these philosophers regarded fossils primarily as curiosities and did not relate them closely to the rocks in which they occurred.

The foundations of stratigraphical geology were laid in the closing years of the eighteenth century by William Smith (1769–1839) when he was resident engineer during the construction of the Kennet-Avon Canal. He was careful to keep separate the fossils which he collected from each stratum exposed in the cuttings of the canal. He noted that each rock layer yielded its own suite of fossils. His next step was to trace in the neighbourhood of Bath the outcrop (or 'course', to use his term) of each rock layer with its distinctive fossil-content. In 1799, urged by some friends whom he had told of his discoveries, he drew up a table of the formation of the Bath neighbourhood, arranged in stratigraphical order and with the distinctive fossils listed. In this area of almost flat-lying and richly fossiliferous rocks he realized that 'the same strata were found always in the same order of superposition and contained the same peculiar fossils'. During the next decade Smith travelled widely whilst pursuing his profession as civil engineer, often covering 10,000 miles in a year either on horseback or in horse-drawn vehicles. He must have had an excellent 'eye for country' and took every opportunity to note the nature and disposition of the rocks and to collect their fossils. In 1815 he published at considerable financial loss the first geological map of England and Wales on the scale of five miles to the inch. In the following year appeared the first parts of his book *Strata Identified by Organized Fossils*, with plates engraved by J. Sowerby. Each plate was of a colour matching that used in his map for the particular formation in which the fossils occurred.

On his map Smith traced what today we would call rock-units or formations, that is groups of beds of distinctive lithology which, if they contained fossils, yielded a distinctive assemblage. Comparison of parts of the succession found in Gloucestershire and Yorkshire showed that though certain beds changed their lithology when traced into Yorkshire, yet they still contained some of Smith's 'peculiar fossils'. Smith named his units or formations in a rather random way. To some he gave names, such as Portland Rock and London Clay, which indicated both the nature of the rock and a type locality. For others he adopted local terms, such as Clunch Clay or Red Rhab, but most of such terms have fallen into disuse.

Smith, by his commonsense methods, found the key to working out the succession of beds. In the early years of the nineteenth century detailed local successions were hammered out by others in many parts of the British Isles. The resulting proliferation of local names soon caused confusion, for as yet methods of recognizing time-equivalents had not been sufficiently developed.

Stages

In the early 1840's a Frenchman, A. D'Orbigny (1802–1857), who had studied in detail the Jurassic and Cretaceous rocks of France, introduced a new concept intended to bring order into this chaos of local successions. He regarded the fauna of the beds as of paramount importance and introduced the term *étage*, now stage, for beds (whatever their lithology) which contained a distinctive faunal assemblage. 'I proceed solely according to the identity in composition of the faunas, or the extinction of genera and families.' The word extinction is significant. For nearly fifty years a bitter controversy had been raging between those who invoked catastrophies of nature to explain most geological phenomena and those who preferred to interpret them by more gradual means. D'Orbigny was a catastrophist and took for his starting-point 'the annihilation of one life-form and its replacement by another'.

Despite this erroneous premise, D'Orbigny's concept of stages, each characterized by its distinctive faunal assemblage, was of fundamental importance. Stages are the major units used in the subdivision of the stratified rocks on a palaeontological basis and hence in correlation.

Zones

Whilst D'Orbigny had been thinking in terms of broad faunal divisions traceable, as he hoped, across the whole earth, other workers had been studying the vertical range of different species of fossils. This work was crystallized by the German A. Oppel (1831–1865) in his concept of zones 'which, through the constant and exclusive occurrence of certain species, mark themselves off from their neighbours as distinct horizons'. Zones are today regarded as the minor units of palaeontological subdivision and correlation. In each area a succession of zones can be recognized in the rocks that make up a stage.

Stages and zones are, therefore, time-stratigraphic units, in the sense that they are major and minor groups of rock deposited during certain periods of geological time. Some prefer to group a number of stages together as a major division of a system. For example, the Lower Cretaceous Series includes the stages Berriasian to Albian shown in the Appendix. In any one area, owing to breaks in the succession resulting from local phases of earth movement followed by erosion, rocks representing the full sequence of stages shown in the Appendix will not be present. The term Age is used for the period of time during which the rocks of a particular stage were deposited in its type area. The relation of geological-time units and time-stratigraphic units is shown below:

Geological-time units	*Time-stratigraphic units*
Era (Mesozoic)	Group (Mesozoic)
Period (Cretaceous)	System (Cretaceous)
Epoch (Lower Cretaceous)	Series (Lower Cretaceous)
Age (Albian)	Stage (Albian)

We can therefore say that during the Albian Age, parts of the British Isles were a land area, undergoing erosion, other parts were submerged beneath a sea in which were deposited the beds of the Albian Stage, and these beds included two formations, the Upper Greensand and the Gault, whose relations will be discussed later.

Zone fossils

The fossils used for recognizing the presence of a particular zone must have a limited time-range, a wide geographical distribution and should, ideally, be easily recognizable. The free-swimming crinoid, *Marsupites testudinarius*, is an almost ideal zone fossil, for owing to its mode of life it is widely distributed in a very limited thickness, usually less than fifty feet, of the Upper Chalk (Senonian Stage) of the British Isles, northern France and Germany .Moreover, with its distinctive ornamentation (see Fig. 33 D) even isolated plates are easily recognizable. Rapidly evolving pelagic types such as certain foraminifera, the ammonoids and the graptolites are other examples of good zonal fossils, though the graptolites had fragile skeletons and are therefore usually preserved only in fine-grained deposits. It must be emphasized that zones are characterized by the occurrence of a particular assemblage of fossils, one of which is chosen as the zone fossil and after which the zone is named. It is not necessary to find the zone fossil to prove that one is dealing with beds of a particular zone. Evidence of the distinctive zonal assemblage is sufficient.

It is not always possible to erect a zonal sequence based on pelagic forms. For example, the fauna of the Carboniferous Limestone of Dinantian Age is almost entirely *benthonic*. Pelagic goniatites do occur, but they are so uncommon that if we relied entirely upon them it would be impossible for practical purposes to zone the beds. We have already referred (p. 50) to the gigantid productids and large solitary corals with prominent axial columns which are characteristic of the highest beds of the Dinantian Stage. At lower levels in the Carboniferous Limestone are to be found other assemblages of corals and brachiopods, and the beds can therefore be subdivided into a series of zones and sub-zones, each with its own distinctive assemblage. The several hundred feet of Upper Chalk underlying the *Marsupites testudinarius* zone often yields specimens of the spatangoid sea urchin *Micraster*. This sea urchin shows steady evolution along a number of lines of minor morphological change (Fig. 37). Even a broken fragment of *Micraster* will usually show, by the degree of projection of its lip, the pattern of its ambulacral areas, etc., how far it has advanced along an evolutionary series. A burrowing sea urchin has therefore proved to be a very satisfactory form for zoning a considerable thickness of the Upper Chalk, but again it is not necessary to find a *Micraster*, for other sea urchins, starfish, lamellibranchs, etc., have limited vertical range in the Upper Chalk. The assemblage of the forms characteristic of the zone of *Micraster cor-angiunum* differs from that found in the underlying zone of *Micraster cor-testudinarium*.

The zonal palaeontologist studies the vertical and horizontal distribution of fossils, and then selects those which are most suitable for his purpose. Macro- and micro-palaeontologists working on the same succession of beds will each have their own zonal sequence. Problems of systematics are to them of very minor importance. For example, the systematic position of the conodonts (Fig. 11) is extremely debatable.

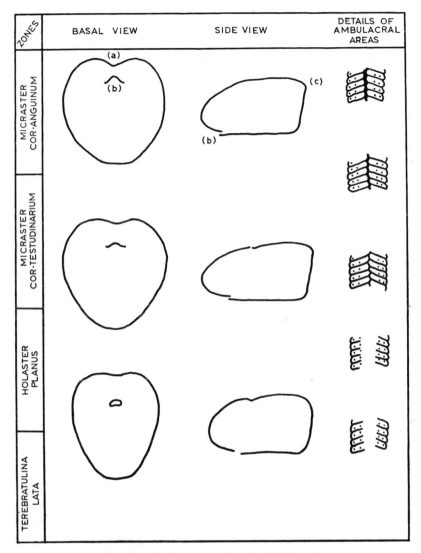

Fig. 37. Evolutionary changes in *Micraster*

This sequence shows progressive change in (a) the depth of the anterior sulcus; (b) the position of the mouth and the projection of the labrum; (c) the height of the posterior carina; and the detail of the ambulacral areas

They have been referred to as the teeth of fishes or of worms, as the radula teeth of gastropods, as the dermal plates or gill arches of fish, or even as unspecified parts of cephalopods or arthropods. The conodonts have been divided into a large number of morphological genera and species. Certain of these which possess a restricted time-range and wide distribution are proving of great value in the correlation of the Palaeozoic rocks, both of Europe and of North America.

Again, the zonal palaeontologist will use whichever group of organisms is most suitable for his purpose. The first attempts to zone the Upper Carboniferous Coal Measures of Westphalian Age rather naturally used plants. But this did not prove too easy in practice. The zones were based on the vertical range of species and it was distinctly uncommon to find plants with a sufficiently limited time-range. Then, less than forty years ago, certain workers began to study the hitherto neglected non-marine lamellibranchs. Not only do these occur abundantly in the so-called 'mussel bands', but it was found that the different species have limited vertical ranges. As a result of discovering a more easily workable zonal sequence, knowledge of Coal Measure stratigraphy advanced enormously, particularly with regard to the age relations of the chief coal seams of the different coal-fields.

Finds of vertebrates are usually too infrequent for them to be used for zoning, except at a few horizons, notably in parts of the Old Red Sandstone (p. 48), in the Beaufort Series of South Africa (p. 56) and in the Lower Tertiary Beds of the western United States (p. 60). In these beds vertebrate remains are not only reasonably common but are almost the only fossils to be found.

The thickness of zones is very variable, for it is determined partly by the rate of sedimentation, but more particularly by the evolutionary changes of the fossils which are used for zoning. The marine clays of the Jurassic, for example, can be divided into a large number of zones, owing to the rapid evolution of the zonal ammonites. Zones in these beds are often measurable in inches. On the other hand, in limestone such as the Carboniferous Limestone and the Chalk, where slowly evolving benthonic forms have to be used, the thickness of individual zones may be measurable in many scores of feet. In poorly fossiliferous rocks, e.g. parts of the Old Red Sandstone, the zonal framework may be even coarser.

Diachronism

Let us now examine the relations of the Upper Greensand with the Gault. At Folkestone, in east Kent, the Chalk, with a basal phosphatic nodule bed indicating a non-sequence, rests on some 130 feet of blue clay, the Gault. The Gault is richly fossiliferous and yields numerous ammonites by which it has been subdivided into a large number of zones. Following the base of the Chalk westwards the basal nodule bed is always present, overlain by beds yielding the same lowest Cenomanian fauna—clearly the base of the Chalk is a time plane. But in west Kent the clays of the Gault begin to be separated from the basal Chalk by a few feet of glauconitic sand and sandstone, the Upper Greensand, which yields the same zonal ammonites as the highest bed of the Gault at Folkestone. In the western end of the Weald, along the Surrey-Hampshire border, the Upper Greensand is much thicker and contains ammonites found at Folkestone in older beds of the Gault. The base of the Upper Greensand is therefore diachronous, for this lithological line is cutting across the time planes shown by the zonal ammonites. In the Isle of Wight and in Dorset the base of the Upper Greensand (see Fig. 38) is not of highest Upper Albian age as in west Kent, but is near the Upper-Middle Albian

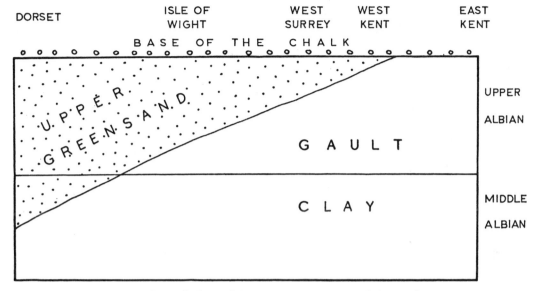

FIG. 38. Relations of the Upper Greensand and the Gault

junction. The particular conditions of sedimentation that gave rise to the lithological unit which we call the Upper Greensand therefore began in late Middle Albian times in Dorset and spread slowly eastwards. This example of the diachronous behaviour of a rock-unit is by no means unique, and must always be borne in mind when one is comparing the detailed succession of beds in different areas. Diachronism is often suspected, but can only be proved (a) when the zonal sequence is sufficiently refined and (b) when the zonal forms have been found at an adequate number of localities.

Facies

Whilst there are sufficient zonal ammonites in common to prove the contemporaneity of the Upper Gault and of the Upper Greensand, the general faunal assemblage of the two lithological units is markedly different. The Gault yields abundant ammonites, belemnites, thin-shelled lamellibranchs, together with rarer solitary corals, crabs, gastropods, etc. In the Upper Greensand of Dorset ammonites are rare, and the fauna consists mainly of thick-shelled oysters and pectiniform lamellibranchs and numerous sponges and echinoids, whilst corals are not uncommon. In the one case we have the fauna of a muddy sea floor, in the other forms characteristic of sandy and probably shallow-water conditions. To use an alternative terminology, the Gault is the clayey facies and the Upper Greensand is the sandy facies of the Upper Albian—the term *facies* (Latin: 'aspect') meaning the lithological and faunal characteristics of beds laid down in a particular environment.

Many fossils, particularly benthonic forms, lived only under certain environmental conditions, that is, they are facies fossils, and as such must have a limited chronological value. Over sixty years ago, when the first zonal scheme for the Albian Beds was put forward, the highest beds of the Upper Greensand were regarded as belonging to the zone of *Aequipecten asper* (Fig. 33). Later, when a more detailed sequence of zones based on ammonites had been worked out for the higher Albian and basal Cenomanian deposits, it was realized that *Aequipecten asper* was a facies fossil. It occurred in beds of 'greensand facies', but in England these are of highest Albian and in north France and Germany of lowest Cenomanian age.

The Ordovician rocks of western and central Wales consist of a thick series of shales and mudstones, zoned by the rapidly evolving pelagic graptolites. Towards the English Border this shale or graptolitic facies passes laterally into more arenaceous beds—siltstones, sandstones, grits with some thin limestones yielding a fauna of benthonic trilobites and brachiopods. Another zonal sequence has been erected based on the vertical distribution of different assemblages of trilobites and brachiopods. Fortunately, at intervals in this 'sandy' facies of the Ordovician there are shales and mudstone bands yielding graptolites, so it is possible to prove the *interdigitation* of the two facies and to correlate the two zonal schemes.

It is particularly in the interpretation of shallow-water deposits laid down on past continental shelves that facies change may raise difficulties. With adequate exposures, or borings, the lateral and vertical extent of the individual facies can be mapped, but unless a zonal sequence, independent of facies, can be both established and traced, it is impossible to prove the time relations of the different facies, their degree of diachronism; also, whilst unconformities and non-sequence can be recognized, the magnitude of these breaks in sedimentation cannot be assessed unless the age of the beds above and below each break is accurately known.

Boundary problems

The geological systems, the major divisions of the stratigraphical column, were founded originally in various parts of north-western Europe. In each type area the system was a clear-cut unit, composed of beds of distinctive lithology and fauna, separated from the beds above and below by unconformities. But as the systems were traced away from their type areas, these unconformities became less and less distinct and finally the beds of one system were overlain with perfect conformity by those of the next, younger system.

In areas where there is no physical break the boundary between the systems must be drawn on palaeontological evidence. This is not the easy matter it would be if D'Orbigny's view of the catastrophic wiping out of the faunas of successive stages were correct! Instead, with greater knowledge, we have found that in areas of continuous deposition there is a transition belt in which the forms characteristic of the underlying stage are dying out and are being replaced by those typical of the younger stage. Moreover, the level at which the new fauna first appears may be different for different groups of fossils. In south-west England, for example, the position of the Devonian/Carboniferous boundary can be drawn at different levels according to whether one relies solely on the change in the trilobites, the goniatites, the corals or the brachiopods. An arbitrary line must therefore be drawn for convenience, but rather naturally its exact position is very controversial. The same difficulties often arise in determining the boundaries of zones.

A rather different problem arises in the case of the Rhaetian Stage. As shown on p. 57, in the type area of western Europe the long continental episode of Permian and Triassic times was ended by an extensive marine transgression. In England the Rhaetic Beds of unusual and very distinctive lithology were laid down during this transgression. Both lithologically and faunally they differ sharply from the underlying Keuper, but pass upwards into the overlying Lower Jurassic strata, the Lias. The Tethyan region, on the other hand, was an area of continuous marine sedimentation. Here the Rhaetian Stage yields numerous ammonites, which are not found in the Rhaetic Beds—indeed, ammonites first appear in Britain some little way up the beds of the Hettangian Stage. Moreover the ammonites of the Rhaetian Stage are Triassic in their affinities, with a major break in the ammonite faunas coming above the Rhaetian Stage. Therefore one can either group the Rhaetic Beds with the Jurassic System on a basis of their behaviour in the British Isles or regard them as an abnormal facies of the Rhaetian Stage, which on the ammonite evidence should be regarded as highest Triassic.

This again emphasizes the difficulty of drawing man-made divisions through rocks which by their lithology and fossil content represent gradual geographical and evolutionary change through periods of time measurable in millions of years.

Unravelling complex structures

The value of zone fossils in unravelling the geological structure of complicated ground is admirably shown by the classic work of the great geologist Charles Lapworth (1842–1920) in the Southern Uplands of Scotland. Before his work the Southern Uplands were regarded as made up of a great thickness (26,000 ft.) of greywackes (ill-sorted sandstones and grits) dipping steadily northwards. At intervals in this conformable succession occurred beds of black shale up to 600 feet thick. The shales contained biserial graptolites, the greywackes very occasionally monograptids. As the same two types of graptolites were repeated throughout this great thickness of rocks, it was assumed (a) that the graptolites were useless as time-indices and (b) that the repetition could be explained by the 'Theory of Colonies'. It was thought that the biserial and the uniserial forms lived as separate colonies in different areas. The opening of communications with the one area would enable the biserials to spread into the Southern Uplands. When this line of communication closed and the other opened, the biserials would be replaced by uniserials. After a time migration routes would change and the biserials reappear.

Lapworth worked on the area for ten years. A most skilled collector, he found better preserved specimens than hitherto and was able to recognize that several different faunas occurred in the black shales. Then in one day's collecting at Dobb's Linn, near Moffat, he found the key to the structure of the area. As he worked across the exposure, he put the first set of graptolites into his right-hand trouser pocket, then successively filled up his other right-hand pockets, and then worked down his left side. That evening he laid out the contents of his pockets. The series in his left-hand pockets closely matched, but in the reverse order, those in his right-hand pockets. Clearly parts of the beds were inverted. He realized that the outcrops of black shales were not different beds, but were always the same bed that had been brought to the surface in the cores of tightly packed isoclinal folds (Fig. 39). Lapworth had also proved that the hitherto despised graptolites were not an unprogressive group; on the contrary they

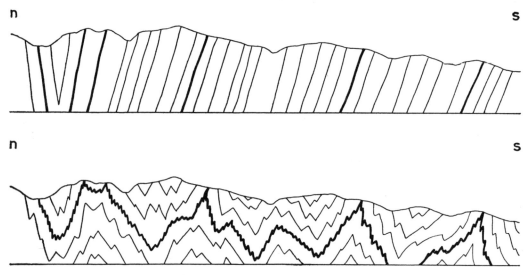

FIG. 39. Two interpretations of the structure of the Southern Uplands of Scotland

Graptolitic shales are shown by the thick black lines. The shales outcrop where these lines cut the profile. In the upper section (the older view) the beds have a uniform dip and several beds of graptolitic shale were believed to be present. Lapworth, however, proved, as is shown in the lower section, that there was only one bed of isoclinally folded graptolitic shales

were a rapidly evolving group (Figs. 17 and 19) which provided the key for unravelling the complicated stratigraphy and structure of the shales of the Lower Palaeozoic, not only in the Southern Uplands, but in all other outcrops.

Another classic piece of work was that of S. S. Buckman (1860–1929) on the oolite limestones of Middle Jurassic age outcropping along the edge of the Cotswold Hills near Cheltenham. At first glance these beds are a conformable series of a wide variety of oolitic limestones. Buckman subdivided the beds in great detail, recognizing a large number of local units of distinctive lithology and with different assemblages of brachiopods, sea urchins and lamellibranchs. Careful comparison of the numerous exposures in the district showed that the Upper Trigonia Grit not only rested on different horizons but that it was usually underlain by a bored surface (Fig. 4). The same relation, though less marked, occurred beneath the Lower Trigonia Grit. Buckman therefore showed

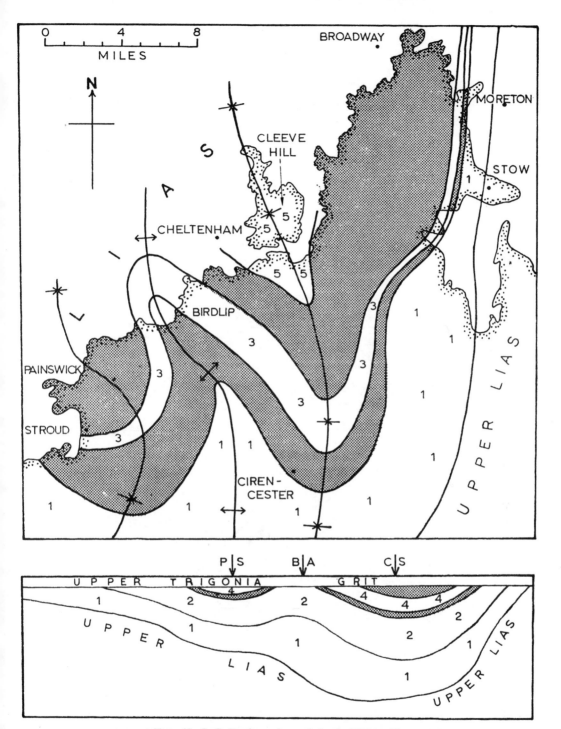

Fig. 40. S. S. Buckman's work in the Cotswolds

In the section below, the Upper Trigonia Grit rests on varying horizons of the underlying beds. Before the Upper Trigonia Grit was deposited, there must have been gentle folding (for the vertical scale is greatly exaggerated) to form the Painswick (P.S.) and Cleeve (C.S.) synclines and the complementary Birdlip Anticline (B.A.) Above is a map showing the beds on which the Upper Trigonia Grit rests. The thick continuous lines with arrows mark the position of the fold axes. The line backed with dots marks the outcrop of the Oolitic limestones on the scarp face of the Cotswold Hills overlooking the Liassic clays of the Vale of Gloucester.

NB The beds on the map and section are numbered in their order of deposition. For clarity, beds 2 and 4 are stippled on the map and beds 3 and 5 on the section

76 THE STUDY OF FOSSILS

that the deposition of these beds was controlled by minor axes of contemporaneous movement, with a complete succession present in the Cleeve Hill syncline, but with periods of non-deposition and erosion along the bordering Birdlip anticline (Fig. 40). His demonstration of the incomplete nature of what appears at first sight to be a conformable and complete succession of beds is fundamental, for it is not an isolated phenomenon, but a possibility that a geologist has always to bear in mind.

The two examples given above may seem to be entirely academic, but this is not so.

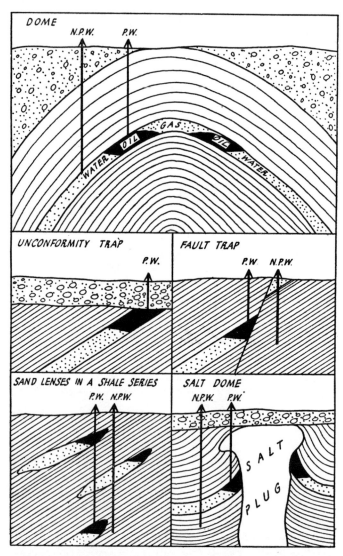

FIG. 41. Some of the structures in which oil may accumulate
Oil accumulations in black; P.W.: productive well; N.P.W.:
non-productive well

Whether or not deposits are of economic importance is in one sense a matter of chance. All sedimentary rocks were laid down under the influence of the same natural laws. The micropalaeontologist engaged in trying to deduce the structure of an oil-field from the specimens obtained from borings attempts to zone the beds penetrated. If he can do so, certain zones will form 'marker horizons' enabling him, by considering their position in three dimensions, to locate 'traps' (Fig. 41) in which oil may be located. The oil geologist is not solely dependent on fossils. In recent years various physical methods have been developed which enable him to recognize slight variations in the lithology of the beds penetrated by a boring, by the differing penetration of gamma rays, neutrons, etc., but these geophysical methods alone cannot determine the geological age of the beds at different depths in the borings. This can only be done if the beds have yielded sufficient fossils. Again, the structure of a coal-field can be thoroughly understood only if sufficient fossil bands can be traced through strata disturbed by faulting and folding.

VI

Fossils as Indicators of Environmental Conditions

THE biologist studying living animal or plant communities is in a very different position from the palaeontologist attempting to reconstruct the ecological conditions of the past. In the first place the fossil material may have been transported after death by the wind, by bottom currents, by floods, or perhaps by other animals, so that the place of fossilization of an organism may be widely separated from its place of life. Secondly the fossil content of a typical marine deposit, such as the Gault Clay (p. 71), combines in the same stratum representatives of the infauna (certain lamellibranchs, echinoids, etc.) which lived in the mud, of the epifauna (crabs, other lamellibranchs and gastropods) which crawled over the mud, and of the nektonic and planktonic forms (ammonites, belemnites, foraminifera, etc.) which inhabited the waters above the sea floor. When one is dealing with extinct forms of life it is often debatable to which of these categories each might have belonged. Furthermore, the palaeontologist has only a partial representation of the life of the past on which to work. This naturally will make deductions about food-chains, life cycles, etc., virtually impossible.

The distribution of living organisms is profoundly influenced by slight differences in temperature, salinity, turbidity and so on. These differences can be measured by the ecologist and their effects assessed, but the palaeontologist is in a much more difficult position. He must attempt to deduce such ecological factors from a study of the sediment in which the fossils are enclosed. During its lithification the character of the sediment may have been altered, perhaps profoundly, by some of the diagenetic changes mentioned on p. 20. With our present knowledge we can recognize major changes in salinity, temperature, etc., only from the character of the sediments. A study of the distribution of particular fossils in relation to the nature of the enclosing sediment may reveal suggestive relationships but fail to interpret them.

A recent and thorough study by G. Y. Craig of a two-foot-thick band of Lower

Carboniferous shale near Kilsyth on the borders of Stirlingshire and Dumbartonshire, Scotland, is a good example of the difficulties of reconstructing palaeoecological conditions. These shales are regarded as shallow subtidal deposits. The lowest four inches of the band are slightly more fissile, more pyritic and of rather finer grain size than the beds above, but these lithological differences are slight. There is, however, a marked change in fauna. The fissility of the basal four inches is due to the abundance of the valves of the pectiniform lamellibranch, *Posidonia corrugata*, lying along the bedding planes. The valves are unworn and show a complete graduation in size from spat to adult. The ostracod, *Waylandella cuneola*, was very probably another indigenous form, for measurement of more than 300 specimens shows several well-defined size groups

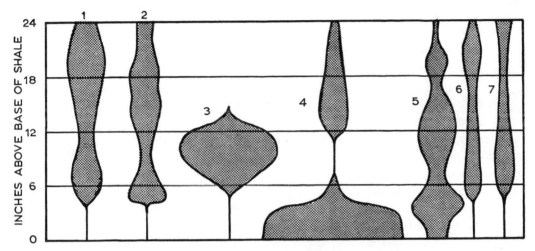

FIG. 42. Palaeoecology of a two-foot band of shale (after Craig)
Width of stippled areas indicates the variation in abundance of the following organisms:
1. *Lingula*; 2. Burrowing lamellibranchs; 3. *Tornquistia polita*; 4. *Posidonia*; 5. *Waylandella*;
6. Aspidobranch gastropods; 7. Rhynchonellid brachiopods

interpreted as moult stages. The microfauna includes the spat of numerous gastropods but no adults, suggesting that when these forms settled on the bottom at the end of the planktonic larval existence they found the environment unfavourable and died. There is an abrupt faunal change above this (Fig. 42): the posidonias are replaced by mud-burrowing lingulas and lamellibranchs. Many of the lingulas are in the position of growth, in burrows at right angles to the bedding planes. Sessile and crawling members of the fauna include brachiopods, lamellibranchs and foraminiferans, whilst the microfauna is more varied, of larger average size and with more species grading from young to adults than in the Posidonia Beds. One brachiopod, *Tornquistia polita*, is represented by many shells, entire but immature, suggesting that the environment into which the spat fell whilst not lethal was not too favourable. The gastropods are rather a difficulty. The majority of them are aspidobranchs, and on current zoological theory these favour a hard substratum. The gastropods of the shale, however, seem to be indigenous, for they show a graduation from spat to adult. Other forms such as fragments of rhynchonellid shells, which are all incomplete, must have been transported by currents from different environments.

The bulk of palaeontological literature deals with the morphological features, the systematic position and the chronological value of fossils. It is only within recent years that many workers have concerned themselves with the difficult problems of palaeoecology. This new development has rather naturally coincided with the much more intensive study of sediments, both recent and fossil. Deduction as to environmental conditions cannot be based on the study of fossils alone, but must take into account all possible lines of evidence as to the conditions of deposition of the beds under investigation.

Life and death assemblages

A phosphatic nodule bed or the infilling of a fissure in limestone containing rolled and abraded fragments of a large variety of fossils are clearly examples of 'death assemblages', for the different organisms present have been swept together after death. In the Beaufort Beds of the Karroo Series of South Africa great numbers of pareiasaurs (Fig. 29 C) have been found. These reptiles occur in mudstones, their skeletons usually complete, back upwards and with the limbs sprawled out, suggesting that they were trapped in the soft mud in which they are preserved and hence are life assemblages. In shales of early Permian age in Texas, cylinders of hardened clay were found lying at right angles to the bedding. The majority of the cylinders were empty, but in some were the remains of lung-fish. The cylinders are interpreted as the burrows dug by lung-fish to aestivate during times of drought. When the waters returned the majority of the lung-fish must have emerged successfully, but a few had died in their burrows. Most of the invertebrates in the Lower Carboniferous Shale described on p. 79 were clearly part of a life assemblage.

If the majority of fossils in a marine stratum are somewhat worn and broken, are all of approximately the same size and are lying aligned in a definite direction (Fig. 43), this is clear evidence of sorting and transportation by bottom currents. It is, however, usually impossible to be sure of the distance travelled.

In the case of marine deposits, only by careful studies such as those mentioned on p. 79 can life assemblages be distinguished from those which may have been introduced from other *biotas*. The fossil content of beds laid down in brackish or freshwater is much less varied. For example, the Lower Eocene (Landenian) rocks underlying south-east London contain an easily traceable horizon, the Woolwich Shell Bed, which was deposited in brackish-water lagoons. Ten feet or so in thickness, its dark-coloured clays are just crowded with shells. But these shells belong almost entirely to four species only: an oyster, two species of cyrenoid lamellibranchs and a turreted gastropod with prominent ornamentation. One is distinctly pleased to find representatives of other species. The freshwater 'mussel bands' of the Coal Measures show a similar monotony of fauna. In such cases it is safe to regard the assemblages as fossil populations which

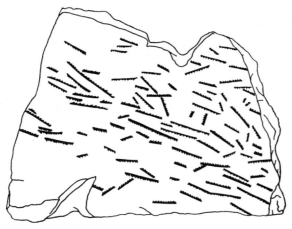

Fig. 43. A bedding plane showing current-aligned monograptids

have suffered a negligible amount of transportation and have had little or no introductions from elsewhere.

'The present is the key to the past'

'The present is the key to the past' is one of the fundamental tenets of geology. If this were not broadly true, the geologist's task would be made very much harder, for, like the old-time catastrophist, he would have to invoke very strange theories to explain the features of rocks. But the tenet must not be applied too rigidly. Scientific study of the processes and changes now affecting the Earth's surface began less than two centuries ago; it is extremely unlikely that this brief span is completely representative of all that has happened during the hundreds of millions of years of geological time. There are certain types of sedimentary rocks, developed at many horizons in different continents, whose exact mode of deposition is very debatable because nothing at all comparable is being formed at present. Moreover, the study has been concentrated on the land areas. Only within the last decade or two have techniques been developed which enable a start to be made on the detailed study of the fauna inhabiting and the deposits now forming on the ocean floors.

We must also remember that in the eyes of the geologist the present is abnormal; the relief of the land areas is unusually pronounced. Many parts of the world were affected in late Tertiary or even Pleistocene times by orogenic movements. The great variations in the extent of the ice-sheets during the Pleistocene glaciations must have caused world-wide fluctuations in the height of sea-level and this would affect the continental shelves and coastal plains. This is very different from much of the geological past, especially during late Palaeozoic, much of Mesozoic and early Tertiary times. There is then little or no evidence of extensive glaciation or of widely developed crustal unrest. The geological evidence strongly suggests that throughout much of Palaeozoic and later times the Earth, in general, was of much lower relief than at present and that the climatic belts were much less pronounced (see p. 94). Knowledge of the Pre-Cambrian rocks is as yet too inadequate to suggest world conditions then.

It is less easy in the organic world to regard the present as the key. According to A. S. Romer, no species of mammal living today can be definitely traced back farther than the Pleistocene; few existing genera beyond the very late Tertiary; few families beyond the Oligocene; whilst the Eocene and particularly the Palaeocene mammalian fauna consist in very large part of orders and sub-orders now completely extinct. The steady widening of the relationship with time of recent and fossil forms applies to the other vertebrates, to the invertebrates and to the plants, though the rate of dissimilarity is not as great as in the rapidly evolving and increasing specialized mammals. Moreover, the environments of modern groups may not be completely representative of those of their close relatives living during the recent geological past. Modern elephants and rhinoceroses are restricted to tropical regions. In the study of the Pleistocene deposits one distinguishes between the 'warm' elephants and rhinoceroses found associated with camels, antelopes, etc., and the 'cold' elephants and rhinoceroses which lived much closer to the ice-sheets, as shown not only by their woolly coats and the type of deposit in which their remains occur but also by the presence in the same bed of the remains of reindeer, musk ox and lemming. Indeed, in these beds the presence of distinctive forms of elephant and rhinoceroses is valuable evidence of climatic conditions.

Change of environment is also known amongst the invertebrates. For instance, the lamellibranchs *Trigonia* (Fig. 31), *Pholadomya* and *Astarte* are common and widespread in the Jurassic rocks, especially in the shallow-water facies laid down in seas whose temperature, as will be shown later, must have been warm temperate. All these genera are living today, but *Astarte* is restricted to boreal regions, *Trigonia* is found only in the warm seas around Australia, whilst *Pholadomya* has retreated to the abyssal regions. The Devonian rocks of New York State are famous for their abundant and well-preserved 'glass sponges' (hexactinellids). The geological evidence indicates that these beds are shelf-sea deposits laid down in water of but moderate depth. Today, however, the hexactinellid sponges have retreated far down the continental slopes and are most numerous at depths between 1000 and 2000 fathoms.

It therefore follows that the older the fauna, the smaller is the percentage of species now living today, and the less reliable the evidence that one can draw from the fossils alone as to the probable environmental conditions. Obviously this problem increases when one is dealing with completely extinct groups, especially if they have no close living relatives.

Reefs

Coral reefs form perhaps the most sharply defined and well-studied biotope today. Fossil reefs have likewise received much attention from palaeontologists and geologists.

Modern coral reefs are biohermal structures built up of the interlocked and encrusted skeletons of madreporarian corals and coralline algae. In and around the reef live a distinctive assemblage of fish, molluscs, echinoids, bryozoans, foraminiferans, etc., which play their part in the formation of the reef by providing calcitic material that helps to infill the interstices of the coral-algal framework. Modern coral reefs, so dependent on the symbiotic relationship between corals and algae, are restricted to warm waters above 18·5°C and to depths of less than 300 feet.

Certain modern coral reefs are of great antiquity. Boreholes sunk to a depth of 4600 feet on Eniwetock Atoll in the Pacific showed that this site had been occupied by shallow-water reef-building corals continuously from Eocene times. Fossil reefs are known from many systems throughout the world, even including the Pre-Cambrian, if the stromatolitic limestones (p. 41) are accepted as of organic origin. In many Palaeozoic examples corals play but a subsidiary part in building the reef framework, their place being taken by other groups such as the extinct stromatoporoids, bryozoans, calcareous sponges, etc. Evidence of the former presence of algae is much more difficult to obtain. Boreholes in recent reefs have shown how prone the porous reef limestone is to diagenetic change, especially to dolomitization, with the obliteration of delicate organic structures, especially those of algae. The same processes must have been active in the past. For example, the Magnesian Limestone of Permian age in north-east England contains reef limestones, built up partly by bryozoans. Recently, evidence of algae has been found in parts of the limestone which are less heavily dolomitized than elsewhere, and it is very probable that algae were the main reef-builders at this horizon. Ramifying through reef limestones of Upper Palaeozoic age are thin layers of calcite, sometimes swelling out into irregular masses. English authors call this 'reef tufa'; American and Belgian workers apply the name *Stromatactis*. It is possible, though not proved, that *Stromatactis* is of organic origin. The beds on the flanks of the reefs may show dips as high as 50 degrees. These depositional dips are greater than the 'angle of repose' of granular material and therefore some sediment binder, most probably *Stromatactis*, must have been present to enable such steeply inclined talus fans to form. The organism must have completely decayed to produce moulds in which calcite was chemically precipitated. In reef limestones, as in many other sediments, our knowledge of the reef-forming organisms is very dependent on the chances of fossilization and of subsequent diagenetic changes.

A distinctive reef assemblage is also present, but its constitution is different from that living on recent reefs. In the Dinantian reefs of the Carboniferous Limestone of the Pennines the reef assemblage comprises many species of brachiopods, often abnormally large, together with gastropods, lamellibranchs and echinoderms. Corals, apart from the solitary rugose genus *Amplexus*, are relatively rare.

As an example of fringing reefs closely comparable to those of the modern Red Sea coasts, we may mention the 'Coral Rag' facies of the Corallian Beds (Oxfordian Stage) of the Upper Jurassic in the Oxford neighbourhood. The 'Coral Rag' is a rubbly unbedded limestone. Many of the blocks are clearly 'coral heads' in the position of growth, with their dead undersides bored by molluscs. The associated reef assemblage consists of certain species of lamellibranchs, including oysters, gastropods and echinoids. The 'Coral Rag' passes laterally into a massively current-bedded limestone, consisting mainly of finely comminuted shell debris, though rolled fragments of the reef assemblage can be found, together with myriads of unbroken shells of small oysters and of certain echinoids. These detrital limestones must have been formed as talus fans on the edge of reefs. In an area to the west of Oxford, where extensive building development provided great numbers of temporary exposures, W. J. Arkell was able to survey,

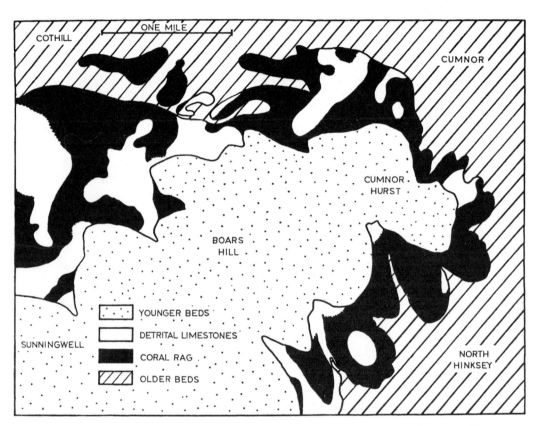

Fig. 44. A Jurassic reef-belt (after Arkell). Geological map of the coral reefs of Corallian age around Boars Hill, Oxford

on the scale of six inches to the mile, the distribution of the two different types of limestone and thus record the shifting plan of the reef-belt. Some of the channels between the reefs remained continuously open, others were closed by the coral growth, whilst yet others opened and broke the continuity of the reef-belt (Fig. 44).

The best-known example of a fossil barrier reef is the Capitan Reef of Upper Permian age on the borders of Texas and New Mexico. Magnificently exposed in the Glass and Guadalupe Mountains and well known from borings for oil elsewhere, it can be traced for nearly 450 miles along the margin of the Delaware Basin (Fig. 45). Owing to the great natural sections and the numerous borings, we have a better knowledge of

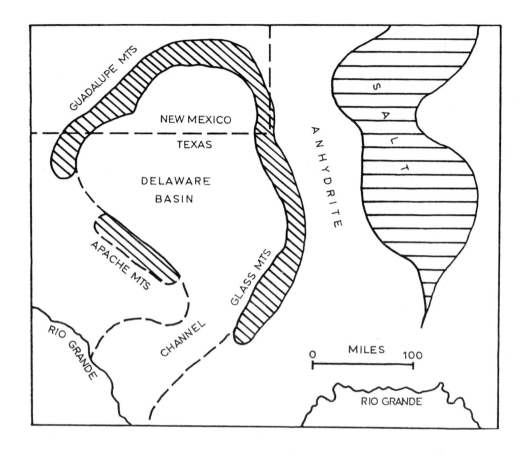
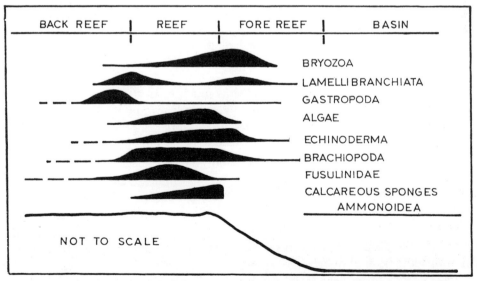

FIG. 45. The Capitan reef-belt (after Newell). On the map the reef-belt is shown by oblique lines, the lagoonal area of evaporites by horizontal lines. Below, diagrammatic cross section across the reef-belt showing the probable distribution of the different groups of organisms

86   THE STUDY OF FOSSILS

the three-dimensional form and changes in character of the Capitan Reef than we have of the Great Barrier Reef off Australia. Only the surface features of the latter have been investigated. The Bikini Atoll is the only modern coral formation which has been penetrated by numerous borings and studied in something approaching the same detail.

In the Middle Permian times patch reefs developed on the edge of the shelf surrounding the basin. Later these patch reefs coalesced to form a massive barrier reef which finally reached a thickness of 1500 to 2000 feet. The reef as it grew upwards also

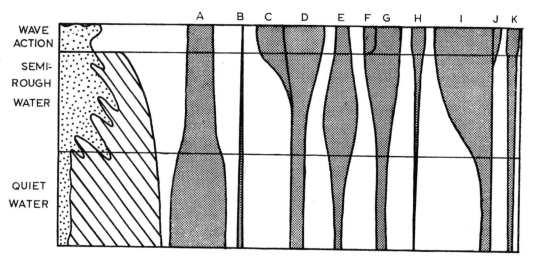

FIG. 46. Changing communities during the upward growth of a Silurian reef (after Lowenstam) Width of the stippled areas indicates the variation in abundance of the following: A *Stromatactis*; B Sponges; C Stromatoporoids; D Corals; E Bryozoans; F Inarticulate brachiopods; G Articulate brachiopods; H Molluscs; I Crinoids; J Blastoids and cystoids; K Trilobites (*Coarse stippled* reef core; *oblique lines* reef flank)

spread basinwards over its own detritus for as much as thirty miles. On the landward side of the reef were extensive lagoons, fed by rivers from higher ground. Evaporation was extremely rapid, so within a few miles of the inner edge of the reef the salinity content of the lagoons was high enough for anhydrite and rock salt to be deposited. The frame-builders of the Captan Reef are many kinds of algae as well as calcareous sponges, bryozoans and hydrocorallines. The reef assemblage comprises many brachiopods, solitary corals, echinoderms and fusulinids. Molluscs were able to survive in the not-too-saline waters of the back reef, whilst the reef-forms extended some distance down the reef front. Indeed the best collecting grounds are on the talus fans of the reef front. In the deep Delaware Basin black bituminous and pyritous limestones and shales with some beds of quartz sandstones were deposited. The presence of benthonic life is indicated by infrequent trails and burrows in beds near the margin of the basin together with occasional nuculoid lamellibranchs. Some of the limestones contain great numbers of isolated spicules of monactinellid sponges. Presumably these sponges grew on the slopes at the edge of the basin. On the decay of the soft parts the spicules would become isolated and may have been carried basinwards by bottom currents. The pelagic fauna is rich in radiolarians and ammonoids. The ammonoids represent complete growth series from juveniles upwards and there is no evidence of current sorting or of secondary concentration of shells after death. The surface waters of the basin may have been rich in phytoplankton. Entire communities of ammonoids may have been killed off by the occasional uprise of the sulphurous bottom waters owing to violent storms or the sliding of sediment basinwards, or by the proliferation of phytoplankton. The stagnant bottom of the Delaware Basin formed an environment very favourable to the formation of petroleum. The weight of the accumulating sediments then forced the oil to migrate into the porous reef limestones.

A fossil archipelago of Silurian age has been investigated in great detail to the south of the Great Lakes in North America. As shown on Fig. 46, it is possible to recognize changes in the biotas as the reefs built up from the bottom to the zone of wave action. The main reef-builders in the quiet-water stage were tabulate corals growing as small bosses and expanded sheets, but as the reef reached rougher waters stromatoporoids played a bigger part. In the zone of breakers the reef assemblage was extremely varied, with certain sessile crinoids showing highly specialized food-gathering structures; the brachiopods, especially the inarticulates, are mainly forms with thick massive shells, whilst amongst the scavenging trilobites the chiton-like streamlined types predominate. In the Devonian rocks of Alberta, Canada, reef limestones have proved reservoirs for oil. Thanks to numerous borings, it is proving possible to reconstruct the horizontal zonation of organisms and hence to infer the direction of waves and currents that moulded these fossil atolls.

Other biotopes

Space does not permit other biotopes, that is, specialized environments with distinctive faunal and floral assemblages, to be discussed at length here. From earlier pages it should be clear that the geologist does not deduce environmental conditions solely from the fossil content of the beds. He pays great attention to the lithological characters of the rocks and especially to the three-dimensional distribution of both fauna and the different rock-types. One line of evidence should supplement and support the other. If it does not, clearly thorough re-examination is necessary. Changes in the composition of the faunas of a particular biotope are much more likely to have occurred than in the type of deposit laid down in that particular environment. The geologist's difficulty is that virtually the same rock may have been deposited in a variety of environments. This applies particularly to the clay rocks. In their interpretation, evidence from the contained fossils is often of the greatest value.

VII
Palaeoclimates

FROM purely lithological evidence it is possible to recognize with certainty only the deposits laid down either under very cold or very arid climates. Boulder clays formed from the melting of ice-sheets are extremely ill-sorted, containing a great variety of rocks, many of which are not of local origin. The boulder clays and their lithified equivalents, tillites, may be underlain by striated and polished surfaces and are sometimes overlain by the closely laminated deposits laid down in *pro-glacial* lakes. Boulder clays pass laterally into outwash gravels and sands, which may show evidence of '*frost heave*' and other *periglacial* features. Evaporite deposits must have been formed from the evaporation of bodies of water under very arid conditions. They are often associated with desert sands, extremely well-graded sands with many of the quartz grains highly rounded and wind-etched. Such sands usually show a characteristic type of current-bedding, from which it is possible to infer the direction of the dominant winds of the time.

But apart from these two extremes, it is very unsafe to make deductions about climatic conditions from lithological features alone. Many different rock-types can form in a variety of environments covering a considerable range of climates. If, however, fossils occur, then the range of possibilities may be greatly reduced and, particularly in the Pleistocene deposits, firm conclusions about climatic conditions may be possible.

Boulder clays often yield fossils, but these have been derived from the beds over which the ice-sheets have passed and clearly are of no climatic significance. Two boulder clays differing in their mass character may, however, be separated by sands, clays or silts. The problem is whether these beds were laid down during an interglacial period, when there was a general major retreat of the ice-sheets and climatic conditions over at least the Northern Hemisphere were broadly similar to those of today, or whether these beds represent an interstadial, just a local oscillation and temporary retreat of the ice-margins. If the beds yield vertebrate remains of a 'warm' character they are clearly interglacial, but if only 'cold' forms are found it is possible that the beds may represent either an interstadial or the beginning or closing phases of an interglacial. The later ice-sheet represented by the overlying boulder clay may well have ploughed off the majority of the unconsolidated deposits lying on the surfaces over which it advanced. That is one of the great difficulties of Pleistocene geology. We can study in great detail the 'Newer Drift', the deposits laid down during the Last Glaciation (the Würm or Weichselian) and during the retreat of the glaciers to their present vastly shrunken limits, but our knowledge of the 'Older Drifts' and of the earlier glaciations is far less complete.

Mammalian remains, for reasons explained in Chapter III, are unfortunately very infrequent. The development of palynology has provided an invaluable method for elucidating Pleistocene and Holocene deposits. Beds yielding pollen and spores are not only more widespread than beds yielding vertebrates but, as the great majority of the plants represented are living today, it is possible to make much more precise deductions about the climatic and environmental requirements of the plant communities. As the climate ameliorates, the area vacated by the ice-sheets will be first occupied by a tundra flora, then by birch and pine, followed by hazel and a mixed-oak forest. Minor oscillations of climate are precisely recorded by the composition of the flora as shown in the table below (after Godwin) of the changes in the past 12,000 years in England and Wales.

Years B.C.	Palynological periods	Vegetation	Climate
500	Sub-Atlantic	Alder—oak—elm birch	mild
3000	Sub-Boreal	Alder—oak—elm—lime	deterioration
5500	Atlantic		climatic optimum
7000	Boreal	Pine—hazel	rapid amelioration
7600	Pre-Boreal	Birch—pine	
8300	Upper Dryas	Park-tundra (birch copses)	cold
	Allerød	Birch woods	milder
	Lower Dryas	Park-tundra (local birch)	cold

Indeed the widely traceable Allerød oscillation is now used as a chronological datum marking the division between late-Glacial and post-Glacial times.

If the deposits between the two boulder clays mentioned above yield a full palynological succession showing the change from arctic → boreal → temperate forest → boreal → arctic, they were clearly laid down in an interglacial, but if the sequence is arctic → boreal → arctic, only an interstadial is indicated.

Moreover, it is now proving possible to correlate interglacials by palynology, for the 'pollen spectrum' of beds during each interglacial has its own distinctive features. In several cases pollen analysis has confirmed tentative correlations made on rather inadequate geological evidence.

F. W. Shotton and his collaborators have recently shown how much can be deduced from really fossiliferous Pleistocene deposits. A number of layers of dark-coloured carbonaceous silts occur in a gravel pit at Upton Warren, Worcestershire. These silts were laid in temporary ponds on the flood plain of a tributary of the River Severn. Vertebrate remains occur in the gravels, whilst the silts have yielded (as well as freshwater molluscs) ostracods, the remains of fish and plants, and a remarkable number of water beetles. The geological evidence that these deposits were laid down during an interstadial of the last (Würm) Glaciation has been confirmed by several radio-carbon determinations, which agree well at an age of 42,000 years. The insects consist of a mixture of 'northern' forms, now restricted to north of latitude 60°N and of 'southern' forms, some of which range well south of this latitude. During an interstadial, climatic conditions would be varying considerably. A rapid amelioration, perhaps comparable to the change from northern Sweden today to that of southern Sweden with its longer and warmer summers, would allow southern species to spread northwards without the immediate extinction of the earlier cold fauna. For a brief period a number of species would be out of phase with the general climatic conditions. The plant assemblage is dominated by a ground flora of biennials and perennials. The rarity of tree and shrub pollen might suggest tundra conditions, but the seeds of a number of temperate plants are present. A possible explanation is that heavy grazing could account for the failure of trees and shrubs to colonize the neighbourhood of the ponds. Bison and mammoth bones occur in the gravels, whilst dung and carcass beetles are very abundant in the pond deposits, suggesting that the ponds were used as drinking and wallowing places. The investigators make the well-justified comment that 'when it is possible to catalogue a fauna and flora as extensively as has been done in this case, problems of climatic and ecological changes become apparent which remain concealed with a more limited approach'.

There is abundant evidence that the climatic stresses of the Pleistocene were preceded by a prolonged period of much more equable climate. Fig. 47 shows the inferred movement during the Tertiary Epoch of the sea temperature of the coldest month (February) along the western coasts of the United States from central Mexico in the south to southern Alaska in the north. The data has been collected by a comparative study of shallow-water molluscan and coral assemblages from various Tertiary horizons with the environments now occupied by their living relatives and with the underlying assumption that the environments of comparable fossil and living associations are similar. The fauna and flora of the British Lower Tertiary Beds similarly indicate subtropical conditions then for areas around latitude 51°N. Seventy-three per cent of the genera of plants found in the London Clay have living relatives today in the Malayan Islands.

For Mesozoic rocks, it is distinctly unwise, if not impossible, to make comparative studies with recent life. But much may be learnt even from extinct groups, for their distribution may be very suggestive. W. J. Arkell made a most thorough world-wide study

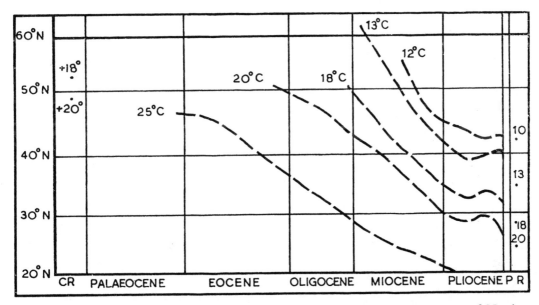

FIG. 47. Change of sea temperatures since Palaeocene times along the west coast of North America (after Durham). The isotherms are for February (coldest month)

of the Jurassic ammonites and of their occurrence in rocks. In Lower Jurassic times the ammonite faunas seem to have been universal (Fig. 48). 'The Hettangian and Sinemurian ammonites of western Canada, northern Alaska, Indonesia and Peru agree at specific level with those of western Europe, and European genera have been recognized in New Zealand, New Caledonia and the Himalayas.' But later there was more differentiation and one can recognize the appearance in the Callovian of a distinctive Boreal fauna of ammonites, which had spread southwards by Lower Oxfordian times in both America and Europe. Then followed a reversal, for in the Upper Oxfordian the Boreal forms retreated before the northward advance of a distinctive assemblage that had developed in the waters of the Tethyan geosyncline. This geosyncline extended roughly along the line of the present Mediterranean to Asia Minor and so eastwards. Reef-building corals are found in the same beds as the Tethyan ammonites. It is noticeable that the Upper Oxfordian coral reefs of the Oxford neighbourhood mentioned on p. 83 yield only six

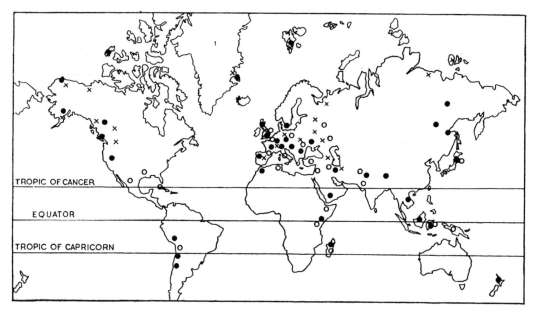

Fig. 48. Climatic provinces shown by the distribution of Jurassic ammonites (after Arkell) *black dots* cosmopolitan Fauna of the Toarcian; *crosses* Boreal cardioceratid Fauna of the Callovian and Lower Oxfordian; *circles* Tethyan perisphinctid Fauna of the Upper Oxfordian

separate species of corals but that the equivalent beds in the Jura Mountains have yielded nearly 200 different species. It is suggested that perhaps only those species with the most highly developed powers of rapid migration and colonization were able to spread from the Tethyan coral belt during the brief periods when conditions were favourable for coral growth in England.

There is no known evidence of glaciation in the Jurassic Beds of the present Arctic or Antarctic region; indeed, floras of a temperate character have been found in east Greenland and Grahamland. The available evidence strongly suggests that there was no ice-cap in the Arctic Ocean during Jurassic times and in fact that its waters were at least as warm as those of the present temperate zones. There must have been some temperature differentiation in Upper Jurassic times between the Boreal and Tethyan realms, but it cannot have approached the sharp latitudinal variation of today.

H. C. Urey and his collaborators have produced a promising method for the absolute determination of palaeotemperatures. Sea-water contains two isotopes of oxygen, O^{16} and O^{18}, whose proportions are believed to be dependent on temperature. The calcium carbonate secreted by organisms will contain these two isotopes and hence record the temperatures of the waters inhabited. Measurements on scrapings of shells of living marine organisms kept in thermostatically controlled baths gave encouraging results, though it was found that certain species grew shell only during a part of the local temperature range and hence did not record the true mean temperature. After investigating a variety of fossil material it was decided that belemnites with their massive guards

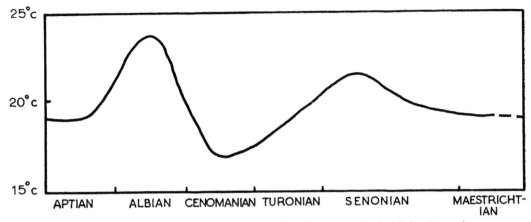

Fig. 49. Temperature variation during the Cretaceous Period (after Bowen)

made up of radiating fibres of calcite were the most promising. Growth rings can be seen in cross-section of the guards and these massive structures are more resistant to diagenetic change than other fossil material. Analysis of belemnite guards from the Cretaceous rocks of Europe (Fig. 49) indicates oscillation of temperature, but the temperatures recorded are within the warm temperate range and are not at variance with the rest of the geological evidence as to the broad similarity of the Jurassic and Cretaceous climates with world-wide equable conditions. A much greater coverage of measurements is necessary before the validity of these palaeotemperatures can be finally assessed. Again, the basic assumption that the ionic composition of the sea-water of the past is similar to that of today is untested, though supported by other lines of geochemical work. These methods do give promise of being able to extend the mapping of the shifting temperature zones of the past back into, at least, the Mesozoic.

In very late Palaeozoic and early Mesozoic times the zonation of temperatures must have been as marked and as great as at present. Unquestioned tillites occur at the base of the Karroo Series of South Africa and of the equivalent beds in Australia, South America, India, etc., whilst in North America, Europe and Asia evaporite deposits and desert sands are widely developed in beds of the same period. We are not primarily concerned here with the problem of the reconstruction of 'Gondwanaland', the land area formed by drawing together the now widely scattered land-masses on which are preserved evidence of this Southern Hemisphere glaciation.

The Theory of Continental Drift was originally produced to account for the Permo-Carboniferous glaciation and the subsequent disruption of Gondwanaland. But drifting of continents may not have been restricted to Gondwanaland. The distribution of the reef limestones of the Jurassic and of the Cretaceous limestones built up by the rudists, an abundant group of giant sessile lamellibranchs, strongly suggests that the contemporary Equator may have passed through the present West Indian and Mediterranean regions. Palaeomagnetism promises to be a method for measuring precisely both the position of the poles during the geological past and also of any movement of the continents, but the results obtained to date are somewhat inconsistent and, moreover, the types of sedimentary rocks whose direction of magnetism can be measured by present techniques are very restricted. As with radio-active dating, considerable advances in techniques are necessary before this new method can be applied widely.

The forest swamps in which the Coal Measures of Europe and the United States were formed have frequently been regarded as growing under tropical conditions. Certainly the giant leaf-fans of the ferns and pteridosperms and the general thinness of the layer of bark suggest subtropical or tropical rain forest, but the narrow leaves of many

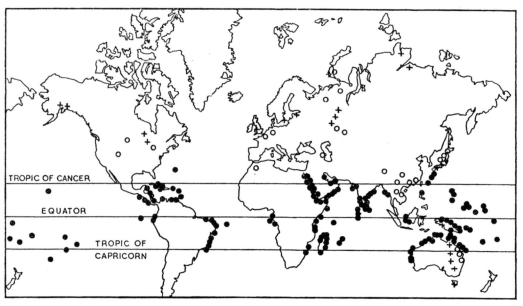

Fig. 50. The distribution of recent and fossil coral reefs (after Termier and Termier)
black dots Recent reefs; *circles* Visean (Lower Carboniferous) reefs; *crosses* Wenlockian and Ludlovian (Silurian) reefs

calamites and lepidophytes have xeromorphic features and it may be that the climate was, at the most, warm temperate. It is impossible to make any deductions about climatic conditions from the meagre traces of the first land plants preserved in the older Devonian rocks.

Although, as shown above, the composition of the reef faunas of the Devonian and Silurian rocks was so different from that of modern reefs that no direct comparison of climatic conditions can safely be made, knowledge of the physico-chemical conditions which today allow calcium carbonate to be secreted in great quantities by organisms suggests very strongly that these Palaeozoic reefs grew in warm waters. If so, the wide latitudinal distribution of these reefs is instructive (Fig. 50) and suggests either that then climatic conditions were much more uniform over the world than they are today or that the relative position of the continents then was markedly different from at present.

VIII

Species in Fossils and Evolution

THE members of a neontologist's species are capable of free and fertile cross-breeding amongst themselves and cannot interbreed with the members of another species. These gene characteristics are reflected in the broad morphological similarity of the members of a species, but there is also, in detail, a certain amount of variation between the individuals of a species. This variation may reach its maximum at the limits of the distribution-area of the species leading to the development of geographical subspecies. The distribution of the species may be controlled by recognizable ecological barriers.

The palaeontologist has no direct evidence of fertility, though when large numbers of individuals are preserved on the same bedding plane it is reasonable to assume that they were members of an interbreeding population. He is entirely dependent on morphological characters and again these are restricted, except most occasionally, to the hard parts which may be far from completely represented. His material may well be inadequate for any satisfactory study of the permissible variation within a species. Moreover, its distribution in the rocks is a function not only of the life habit but also of events that may have occurred between the death and fossilization of an organism. The neontologist in determining the limits of his species is concerned with a single moment of time, the present. The palaeontologist, on the other hand, is concerned, and very vitally concerned, with the effect of time. By the study of the fossil content of successive layers of rocks he can demonstrate the reality of evolutionary change, although, owing to the nature of his material, he is unable to be dogmatic about the mechanism of evolution.

Morphological species

In the early days of palaeontology, species were based entirely on visible morphological characteristics. A specimen showing a certain combination of morphological features was held to be sufficiently distinct from another somewhat similar specimen for the two to be regarded as belonging to different species. Morphological species still have to be recognized today in beds whose fossil content is widely scattered both vertically and horizontally. There is naturally a strong subjective element in deciding on the limits of morphological species. Some palaeontologists, often referred to as 'splitters', lay great stress on very small changes and would recognize numerous species in material which a 'lumper' would regard as belonging to only one species with a permitted amount of variation between individuals.

Lineages

With more intensive bed-by-bed collecting it became possible to arrange certain species in evolutionary series; for example, Rowe's *Micraster* series (Fig. 37), showing a sequence of changes affecting certain morphological features.

The plexus

With the development of biometric studies and the discovery of certain horizons containing great numbers of individuals—in other words, fossil populations—it became possible to study both the limits of variation within a species and also, in much greater thoroughness, changes with time.

Biometric analysis also gave a means of checking the validity of the morphological species. For example, measurements were made of the length, breadth and height of certain brachiopods which occur in great numbers at some localities in the Carboniferous Limestone. Naturally, only specimens which showed no signs of distortion during or after fossilization could be used. A frequency diagram of, say, the ratio height/breadth was then plotted. If the curve obtained was unimodal it showed the normal variation to be expected within a species, but if the curve showed marked signs of bimodality, then either the members of two species had been lumped together, or in that particular bed two distinct species were in the process of evolving from a parent species.

In certain strictly limited parts of the geological column it is possible to find a number of closely spaced beds, each yielding fossil populations. Examples are the 'oyster beds' in parts of the basal Jurassic strata of England and Scotland. From these beds a number of morphological species have been recognized, each characterized by differences in the shape of the attached (left) valve and, in particular, in the relative size and position of the area of attachment (Fig. 51). A. E. Trueman and H. H. Swinnerton have shown that at any one horizon there is a wide variation in the characters of the oysters present and that this variation incorporates the members of several morphological species. They therefore regard these species as capable of interbreeding, for at any one level the variation-curve of the *Ostrea-Gryphea* population present is unimodal. Further, as successive horizons are compared (Fig. 52) there is a steady shift with time of the variation-mode. To quote Swinnerton:

'The evolution of a species must therefore not be thought of as proceeding along a single line leading from one variety, or even morphological species, to another, but as travelling along an intricate plexus made up of a long succession of communities in which

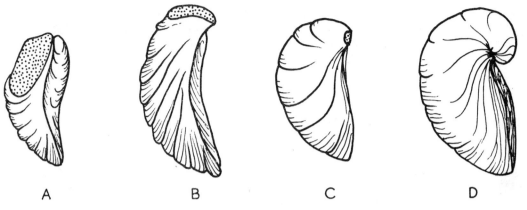

Fig. 51. Members of the *Ostrea–Gryphea* lineage (after Trueman)
A *Ostrea irregularis*; B *Gryphea dumortieri*; C *Gryphea obliquata*; D *Gryphea incurva*. Note the increase in the incurvature of the umbo and the reduction in the area of attachment (stippled)

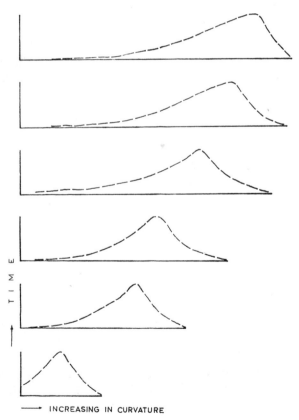

Fig. 52. Variation curves of oyster populations at different stratigraphical levels

G

one or several features pass progressively from a state of incipience to that of dominance. Each community embodies many freely interbreeding varieties, some so distinctive as to be worthy of elevation to the rank of morphological species. In such a plexus each knot represents a mating pair and each thread an individual life history.'

The Coal Measure lamellibranchs

Other examples of fossil interbreeding populations are given by the 'mussel-bands' of the Coal Measures. Detailed investigations have been carried out on the considerable amount of variation shown by the members of any one horizon (Fig. 53). The results have been expressed in pictograms, scatter diagrams and various other statistical methods. Clearly the populations form part of a plexus, and if a number of mussel bands can be found at closely spaced horizons then it is possible to follow through the changes

FIG. 53. Variation in a community of Coal Measure lamellibranchs (after Leitch). The specimens of *Anthracomya salteri* were all collected from one horizon at Rawyards, Lanarkshire

undergone by the plexus. But these changes may be in part due to varying environmental conditions. R. M. C. Eager, in an interesting study of the variations in the shell shape of Recent *unios* and of certain anthracosias from the Coal Measures, has shown that in the Recent unios there is a degree of correlation of the variation of shell shape with ecological station. Arcuate forms, or those with curved dorsal margin and straight or reflected lower borders, appear to be characteristic of relatively swift-moving water, whilst in slowly moving waters there is a greater tendency for shells to have a curved lower border. There is also an increase in relative height and obesity in forms living in more slowly moving waters. A similar study of the anthracosias found at various levels in three Coal Measure cyclothems indicates similar changes in shell form, which appear to be related to the varying grain size of the enclosing sediment; and this must be roughly proportional to the velocity of the water in which the sediment was deposited.

The question of the delimitation of species amongst these highly variable Coal Measure lamellibranchs raises many difficult problems. The geologist is concerned very largely with the time element. He wishes to be able to trace the evolutionary changes in a plexus with time and also to be able to use these forms for zonal purposes and for the precise identification of 'marker' horizons. At one time he may be dealing with a large collection from a particular locality, a collection showing considerable contemporaneous variation and susceptible to biometric analysis. At another time he may have to date a small collection of rather indifferently preserved shells from a borehole. Present experience suggests that the most satisfactory method is to erect morphological species based on holotypes, with as complete statistical information as possible on permissible variation from the holotype within the species. As knowledge increases, so it is possible to become sure of the chronological position of these morphological species in the plexus and hence of their significance in *phylogeny*.

It follows that for most other groups of fossils which are not found as populations the palaeontologist must also think in terms of morphological species. Many of these are founded on only a partial knowledge of the skeleton. For example, only the head shields of many species of trilobites are known. With such scattered and dismembered material, the tracing of phyletic relationships is naturally rather a difficult matter.

The evolution of the Horse

Perhaps the most fully documented evolutionary history is that of the Horse. The essential features have been known for many years, but as more and more material has been discovered it has become clear that the story is vastly more complex than was originally thought. Instead of being a simple line, there have been many unsuccessful offshoots and side branches.

The forerunners of the horses were the condylarths, generalized mammals found in the Palaeocene and Eocene rocks. The early condylarths show a combination of the characters of both carnivores and hoofed herbivores (ungulates), but the later condylarths showed more definite ungulate features. *Phenacodus* (Fig. 35 A) was rather dog-like. The feet had the primitive number of five toes, though the third (middle) toe was stoutest and the first and fifth much smaller. The feet were padded, though in *Phenacodus* each toe carried a small hoof. The wrist and heel were carried slightly off the ground, and as the upper and lower long bones were of equal length, the condylarths were probably not very fleet runners. The skull was very primitive, with eyes of moderate size set in the middle. The brain was both small and simple. The forty-four teeth were arranged in an almost continuous series without any pronounced gaps or diastemata. The cheek teeth were low crowned and with few cusps, suggesting that the condylarths were omnivorous, rather than completely herbivorous, feeders.

Eohippus, the dawn horse, is known from hundreds if not thousands of specimens, usually fragmentary, which have been found in the early Eocene Beds of North America. A number of species have been recognized, varying in size in the adult from about ten to twenty inches at the shoulder. The larger forms were about half the size of a Shetland pony.

Eohippus shows a combination of primitive condylarth characters and of incipient specializations. The back was arched and flexible, the hindquarters high, giving the animals a rabbit-like rather than a horse-like appearance. The long bones of the feet were relatively longer than in the condylarths. The foot was padded, but there were only four hoofed toes on the front foot and three on the hind foot (Fig. 54). In each case the third toe was the longest. The skull was still very condylarth-like, but there was the beginnings of a gap between the incisors and the cheek teeth. The cusps on the molars were starting to form a crest and the dentition is definitely better adapted for dealing with an herbivorous diet, though, as the teeth were very low crowned, *Eohippus* must have browsed on succulent plants.

The early Oligocene rocks of south Dakota have yielded a wealth of remains of *Mesohippus*, the earliest-known three-toed horse. A number of the skeletons are almost complete. *Mesohippus* stood about two feet high at the shoulder and was definitely horse-like. The back was still somewhat arched, whilst the legs were long and slender,

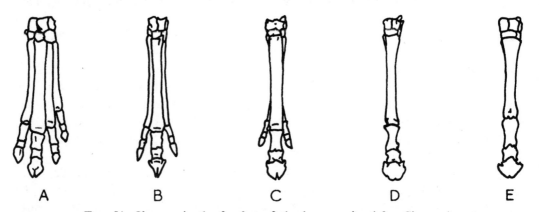

FIG. 54. Changes in the forefeet of the horse series (after Simpson)
A *Eohippus*, Eocene; B *Mesohippus*, Oligocene; C *Merychippus*, Miocene; D *Pliohippus*, Pliocene; E *Equus*, Recent

especially the parts below the elbow and knee. It was clearly a fast runner. Both hind- and fore-feet had three functional toes with a pad between and behind them. The eye was set fairly far back in the skull, whilst the brain showed a most marked advance in the size and complexity of the cerebral hemispheres (Fig. 55). The low-crowned browsing teeth showed two advances. First the premolars closely resembled the molars, thus providing a battery of six grinding and crushing cheek teeth, and, second, the crests on the teeth were now sharp and continuous. *Mesohippus* of the Lower Oligocene grades imperceptibly into *Miohippus* of the Upper Oligocene.

The story now begins to become more complicated (Fig. 56). The anchitheres, best known from the Miocene Beds of Eurasia, are the culmination of the line of multi-toed browsers. *Archaeohippus*, the pygmy horse of the Miocene, is an offshoot, as is the line that leads eventually to *Equus*.

As already mentioned on p. 60, during the Miocene Period there were major changes in environmental conditions in North America. The climate became more arid, with the extensive development of grass steppes in place of lush forests. The blades and seeds of grass contain silica and are often coated with dust and grit. The low simple teeth of the browsing horses would soon be worn out on such a diet. The sequence *Miohippus–Parahippus–Merychippus* shows a major change in dentition. The low-crowned (brachydont) teeth become high-crowned (hypsodont) and an increase in the complexity of the cheek teeth produced an excellent grinder as the lower jaws moved sideways against the upper jaws, whilst most important was the development of bone-like cement filling in the pockets between the deep ridges of hard enamel. The development of teeth able to stand up to a life of grazing was a major advance, but it was not accompanied by such obvious changes in other parts of the skeleton. This independence of the rate of evolutionary change in different skeletal elements is a feature that is not peculiar

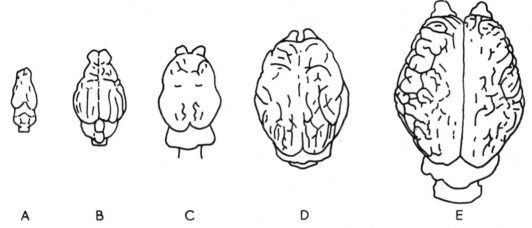

FIG. 55. Changes in the brains of the horse series (after Simpson)
A *Eohippus*, Eocene; B *Mesohippus*, Oligocene; C *Merychippus*, Miocene; D *Pliohippus*, Pliocene; E *Equus*, Recent. Note the increase in size of the brain as a whole and especially of the proportions and convolutions of the cerebral hemispheres

to the horses, but has been noted in many other evolutionary sequences, both of invertebrates and vertebrates.

Merychippus showed an increase in size, some species standing forty inches at the shoulder, that is, as large as many living ponies. The eye was set far back in the skull, the jaws were markedly deeper to accommodate the deep-crowned teeth. The feet were still three-toed, but in advanced species the side toes were short and small and the primitive foot pad was lost. In the forelimbs the ulna fused with the radius, whilst in the hindlimbs the fibula was reduced to a mere spike on the tibia. The effect of these changes was to convert the leg into a rigid weight-carrying structure, moving only in a fore-and-aft plane, and incapable of the limited amount of rotation possible in the earlier Tertiary horses. *Merychippus* was clearly well adapted to wander far and fast over the grassy plains.

By late Miocene times the grazing horses had split into at least six different lines (Fig. 56). As in the comparable split of the browsing horses at the beginning of the Miocene, a number of these lines showed little further change. *Nannihippus* was a

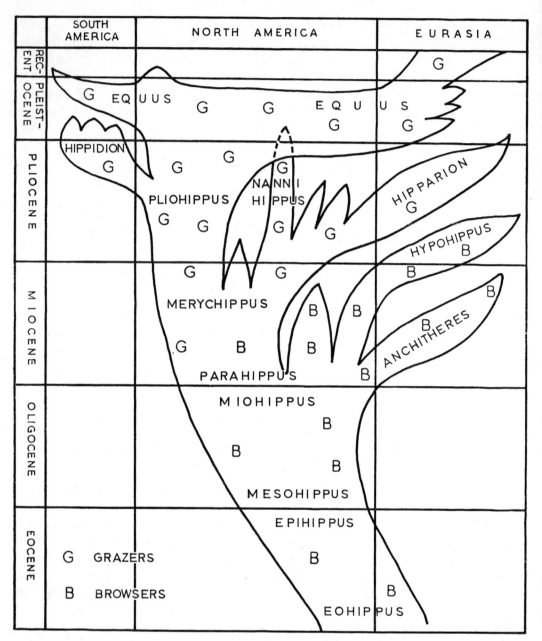

Fig. 56. Evolution of the horses (after Simpson)

pigmy form. *Hipparion* a successful group in the sense that its remains are widely distributed in the Pliocene rocks of Eurasia, and it was the first horse as far as is known to enter Africa.

From an evolutionary standpoint the most important of the groups arising from

Merychippus was *Pliohippus*, the first one-toed horse, which indeed (apart from minor details) was little different from *Equus*.

The evolution of the horses summarized above is much more complex than the straight-line evolution suggested by an over-simplified diagram such as Fig. 55. There was not a steady simultaneous progression in all kinds of horses. For example, during Pliocene times there were in existence three-toed browsing horses (*Hypohippus*), three-toed grazers (*Hipparion*, etc.) and one-toed grazers (*Pliohippus*), and each genus was represented by a number of species. Further, the major advances in brain, teeth and feet were not simultaneous. In the brain the main development occurred early, between *Eohippus* and *Mesohippus*; in teeth considerably later, between *Parahippus* and *Merychippus*; in the feet it is more difficult to be so definite, for the change from the later forms of *Merychippus* with very much reduced side toes to the one-toed *Pliohippus* was a slight one—not really as great an advance as was shown by the earlier forms of *Merychippus*.

Horizontal variation

As well as variation with time, the palaeontologist can sometimes study horizontal variation. This is possible only in beds of relatively uniform lithology (otherwise changes due to ecological conditions become too important) which can be traced for a considerable distance. These beds must also have been laid down in as narrow a time interval as possible. If the beds are diachronous, then another factor, that of evolutionary change with time, will be involved and this will not affect all the beds to the same extent.

The brachiopod faunas of certain horizons of the Jurassic and Cretaceous rocks of Britain have proved very suitable for such studies. Brachiopods are sessile animals; they are often found in the rocks in colonies, and their shape variation can be easily studied by biometric methods. Further, the stratigraphical palaeontology of the Jurassic and Cretaceous rocks is known in great detail, so that any time-element can be assessed with precision.

W. S. McKerrow, who studied several thousand terebratulids—collected at intervals along forty miles of outcrop of the Fuller's Earth Rock (Bathonian Stage) of the Middle Jurassics—found that while there was a considerable variation-range at any one locality, yet there were some overall lateral changes (Fig. 57). These changes presumably reflect

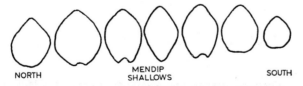

FIG. 57. Horizontal variation in brachiopods of the Fuller's Earth Rock (after McKerrow). The specimens illustrated are members of the same species of ornithellid brachiopod collected from localities along 40 miles of outcrop

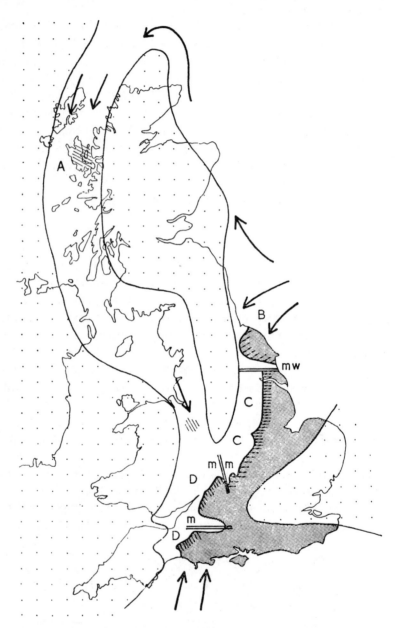

FIG. 58. Faunal provinces and migration routes of Middle Liassic (Domerian) brachiopods (after Ager)

Probable land areas dotted; *outcrops of Middle Lias* lined; *probable subsurface extent* stippled; *migration routes* arrows; *belts of shallows* double lines; m—Mendips; mm—Vale of Moreton; mw—Market Weighton. The faunal provinces recognized are A Hebrides; B Yorkshire; C Midland; D South-western

minor variations in environmental conditions and it is to be noted that such variations are arranged symmetrically about the Mendip Hills. There are many lines of purely geological evidence indicating that in Lower and Middle Jurassic times the neighbourhood of the present Mendip Hills was a belt of relative 'shallows' as compared with the areas to the north and south.

D. V. Ager, by an examination of the rhynchonelloids and terebratuloids of the zone of *Pleuroceras spinatum* of the Middle Lias (Domerian Stage) of Britain, has been able to recognize a number of faunal provinces and subprovinces, based on the species present and their relative abundance. Brachiopods occur most infrequently in the underlying beds, so during *spinatum* times a rich and varied fauna must have spread into the British Isles. As shown on Fig. 58, Ager recognizes three distinct migration routes, and it is interesting that the spread of certain species seems to have been stopped by the three belts of 'shallows' shown.

In the case of certain Cretaceous brachiopods, notably those of the Lower Greensand (Aptian Stage), it is possible to recognize in some species the development of varieties or geographical subspecies which have a restricted horizontal distribution.

* * * * *

It will be seen from this chapter that when the palaeontologist has well-preserved and abundant material to study he can detect variation in populations and to some extent minor modifications due to ecological conditions and geographical subspeciation. But the main difference between the outlook of the palaeontologist and the neontologist is that the former is always very conscious of the possibility of the effects of time. It has been said that 'Palaeontology is the handmaiden of Stratigraphy'. Certainly Earth history is the central theme of geology. The study of fossils is of the greatest importance in unravelling the complexities of Earth history, especially in the Phanerozoic rocks laid down during the last 600,000,000 years. Fossils provide the chronological framework without which it would be impossible to piece together the evidence obtained from the various methods of studying rocks and so provide a connected and coherent account of Earth history.

APPENDIX

The Chief Divisions of the Geological Column

PHANEROZOIC

Group	System	Series	Marine stages of Western Europe	Main rock-units in Great Britain
Quaternary	Pleistocene	—	Monasterian Tyrrhenian Sicilian Calabrian	Boyn Hill Terrace Red Crag
Tertiary	Pliocene	—	Astian Plaisancian	Coralline Crag —
	Miocene	—	Pontian Vindobonian Burdigalian Aquitanian	Absent
	Oligocene	—	Stampian Tongrian	Beds of the Isle of Wight
	Eocene	—	Bartonian Auversian Lutetian Ypresian Landenian Montian	Barton Beds Bracklesham Beds London Clay Lower London Tertiaries —

APPENDIX

Group	System	Series	Marine stages of Western Europe	Main rock-units in Great Britain
Mesozoic	Cretaceous	Upper	Danian	—
			Maestrichtian	Chalk
			Senonian	
			Turonian	
			Cenomanian	
		Lower	Albian	Upper Greensand/Gault
			Aptian	Lower Greensand
			Barremian	Wealden
			Hauterivian	
			Valanginian	
			Berriasian	
	Jurassic	Upper		Purbeck Beds
			Portlandian	Portland Beds
			Kimeridgian	Kimeridge Clay
			Oxfordian	Corallian and U. Oxford Clay
			Callovian	L. Oxford Clay and Upper Cornbrash
		Middle	Bathonian	Great Oolite
			Bajocian	Interior Oolite
		Lower	Toarcian	Upper Lias
			Pliensbachian	Middle and Lower Lias
			Sinemurian	
			Hettangian	
	Trias	Upper	Rhaetian	Rhaetic
			Norian	Keuper
			Carnian	
		Middle	Landinian	(Muschelkalk)
		Lower	Anisian	Bunter
			Scythian	

107

APPENDIX

Group	System	Series	Marine stages of Western Europe	Main rock-units in Great Britain
Palaeozoic	Permian	Upper	Kazanian / Kungurian	Magnesian Limestone
		Lower	Artinskian / Sakmarian	
	Carboniferous	Upper	Stephanian / Westphalian / Namurian	Coal Measures / Millstone Grit
		Lower	Visean / Tournaisian	Carboniferous Limestone
	Devonian	Upper	Famennian / Frasnian	Old Red Sandstone
		Middle	Givetian / Eifelian	
		Lower	Emsian / Siegenian / Gedinnian	
	Silurian	Ludlow		Upper Ludlow Shales / Aymestry Limestone / Lower Ludlow Shales
		Wenlock		Wenlock Limestone / Wenlock Shales / Woolhope Limestone
		Llandovery		Llandovery Sandstone
	Ordovician	Ashgill / Caradoc		Bala
		Llandeilo / Llanvirn / Arenig		
		Tremadoc		Shineton Shales of Shropshire
	Cambrian	Upper		Lingulella Flags of North Wales
		Middle		Menevian Shales of North Wales
		Lower		Comley Limestone of Shropshire / Lower part of Durness Limestone of N.W. Scotland (Both limestones are underlain by sandstones and these by pebbly quartzites)

PRE-PHANEROZOIC

Further Reading

General Geology

HOLMES, A. *Principles of Physical Geology* (1288 pp.) London (Nelson) 2nd Edition 1965

KIRKALDY, J. F. *General Principles of Geology* (327 pp.) London (Hutchinson) 2nd revised Edition 1962

Historical Geology

KUMMEL, B. *History of the Earth* (610 pp.) London (W. H. Freeman) 1961
Well illustrated American textbook, giving an excellent account of world stratigraphy, with much attention to the fossil content of the rocks.

WELLS, A. K., and KIRKALDY, J. F. *Outlines of Historical Geology* (503 pp.) London (T. Murby) 6th Edition 1968
Geological history of the British Isles. Numerous line drawings.

Palaeontology

ANDREWS, H. N. *Studies in Palaeobotany* (487 pp.) London (Wiley) 1961
Well illustrated and interesting, with brief sections on palynology and on techniques used in the study of fossil plants.

COLBERT, E. H. *Evolution of the Vertebrates* (479 pp.) London (Wiley) 1955
Well illustrated and pleasantly written authoritative account.

JONES, D. J. *Introduction to Microfossils* (406 pp.) New York (Harper) 1956
Clearly illustrated account of all types of microfossils and of their use to the geologist.

MOORE, R. C., LALICKER, C. G., and FISCHER, A. G. *Invertebrate Fossils* (765 pp.) London (McGraw-Hill) 1952
Comprehensive, authoritative, well illustrated; emphasis naturally on American forms.

A. MORLEY DAVIES *An Introduction to Palaeontology* (322 pp.) London (Murby) 3rd Edition—revised by C. J. Stubblefield 1961
Describes selected examples of each of the main groups of fossils. Chapters on collection and preservation of fossils and on rules of nomenclature.

ROMER, A. S. *Vertebrate Palaeontology* (418 pp.) Chicago (University of Chicago Press) 3rd Edition 1966
The standard work.

SIMPSON, G. G. *Horses* (247 pp.) New York (Oxford University Press) 1951
Full account of both recent and fossil horses.

Special Topics

AGER, D. V. *Principles of Palaeoecology* (352 pp.) London (McGraw-Hill) 1965

CAIN, A. J. *Animal Species and Their Evolution* (190 pp.) London (Hutchinson University Library) 1956

NAIRN, A. E. M. (Ed.) *Descriptive Palaeoclimatology* (380 pp.) New York (Interscience Publishers) 1961
Critical reviews by experts on the various kinds of evidence that have been used for deducing climatic conditions of the past.

ZEUNER, F. F. *Dating the Past* (516 pp.) London (Methuen) 4th Edition 1948
Primarily concerned with Pleistocene chronology, but contains sections on the various dating methods and on biological evolution and time. Comprehensive bibliography.

Publications of the British Museum (Natural History)

A series of guides costing only a few shillings each, written by experts and well illustrated.

BRITISH CAENOZOIC FOSSILS (132 pp.) 3rd Edition 1968

BRITISH MESOZOIC FOSSILS (207 pp.) 3rd Edition 1967

BRITISH PALAEOZOIC FOSSILS (198 pp.) 3rd Edition 1969

LE GROS CLARK, W. E. *History of the Primates* (127 pp.) 9th Edition 1965

OAKLEY, K. P. *Man, the Tool-maker* (98 pp.) 5th Edition 1967

OAKLEY, K. P., and MUIR-WOOD, H. M. *The Succession of Life through Geological Time* (94 pp.) 7th Edition 1967

SWINTON, W. E. *Fossil Amphibians and Reptiles* (133 pp.) 5th Edition 1967

SWINTON, W. E. *Fossil Birds* (63 pp.) 2nd Edition 1965

Other Geological Guides

British Regional Geology. A series of twenty Regional Guides (North Wales, Wealden District, etc.) published by H. M. Stationery Office for the Geological Survey—price five shillings and upwards each. Description of the geology of each region, often with figures of characteristic fossils.

Geologists' Association Guides (in the course of publication). Each Guide—price two shillings and upwards—contains a number of detailed itineraries giving the best exposures, viewpoints, etc., for studying the geology of a particular neighbourhood. Details of fossiliferous localities. (Obtainable from Messrs Benham and Company, Sheepen Road, Colchester, Essex)

Glossary

ambulacral—Five radiating areas on the test or exoskeleton of an echinoid. They are perforated for the passage of the tube feet (see Fig. 31 A, p. 57, and Fig. 37, p. 69).

anaerobic—Deficient in oxygen or able to live in such conditions.

batholith—A large body of igneous rock, many square miles in area, and thousands of feet in thickness.

benthonic—Sessile, burrowing or creeping forms living on the sea floor.

biohermal—A rock formed very largely of material of organic origin, e.g. a coral reef, or Fig. 9 C, p. 22.

biota—The organisms that live in a specific biotope (q.v.).

biotope—The specific environment of organisms. In the Pleistocene one can recognize a number of biotopes, e.g. tundra, temperate forest, etc., each characterized by a distinctive assemblage of animals and plants.

bryozoa—Colonial aquatic organisms (the moss animals), also known as polyzoa. The members of this phylum form delicate encrusting calcareous colonies with minute apertures through which the individual bryozoans protrude (see Fig. 28 A, p. 54).

cellulose peel—A method of preparing thin sections of petrifactions, particularly useful with fossil plants. A polished surface is etched with acid so that the mineralized cell walls stand out in relief. The surface is then coated with cellulose acetate, which can be peeled off with the cell walls, etc., embedded.

chitin—A horny organic substance which forms the skeleton of certain invertebrates.

country rock—The rock into which mineral veins or igneous bodies have been intruded.

facial suture—A suture on the head shield or carapace of trilobites, separating the free and fixed cheeks and believed to be an aid to periodic moulting of the exoskeleton (see Fig. 16 D, p. 43).

facies—See p. 71 and *interdigitation* (below).

frost heave—When water in the soil and subsoil freezes it must expand. This expansion may heave or in other ways distort the upper layers of the ground.

ganister—A sandstone penetrated by rootlets and with a very low content of alkalis. Ganisters underlie coal seams and are fossil soils.

guard—All that is normally found of the internal skeleton of belemnites. Robust, cigar-shaped (see Fig. 30 C, p. 56), built up of radiating fibres of calcite.

hominoid—The Super family *Hominoidea* includes Man and his nearest relatives, the anthropoid apes. The hominoid australopithecines are very close to the divergence of Man and apes. Some of their skeletal features are more anthropoid than human, others are more human than anthropoid.

horizon—A thin bed recognizable by some distinctive feature, usually its contained fossils.

humic—Carbonaceous rocks in which the organic matter is of a woody nature.

interdigitation—Occurs when two beds of different characters (facies) interfinger with each other, proving (a) that they are of the same age and (b) that conditions of deposition were changing repeatedly from one kind to the other.

orogenic—Mountain building.

palynology—The study of fossil spores and pollen. A recent development of great value in studying the stratigraphy and ecology of the Coal Measures and of Pleistocene and Holocene strata.

pelmatozoan—Echinoderms, such as crinoids (sea lilies), which are fixed in the adult stage to the sea floor (see Fig. 28 F, p. 54).

periglacial—The zone of severe climatic conditions surrounding a glaciated region. 'Freeze-thaw' causes considerable contortion and disturbance of the top few feet of the ground (see *frost heave*).

phylogeny—The evolutionary history of a group of animals and plants.

pro-glacial—In front of a glacier.

seat earth—The fossil soil in which grew the plants that formed the overlying coal seam (see Fig. 27, p. 53).

serial grinding—A smooth face is ground across a fossil, and the structures visible are either carefully drawn or photographed. A known thickness of the fossil is then ground away and the new face drawn or photographed. From the series of drawings or photographs thus made, the structures can be reconstructed in three dimensions.

stolon—Slender chitinous thread embedded (see Fig. 14 C, p. 38) in the centre of the stipes (branches) of graptolites. The thecae appear to originate from the stolon. A comparable structure is found in the living pterobranch, *Rhabdopleura*.

taiga—The belt of coniferous forest between the barren tundra to the north and temperate deciduous forest to the south.

tanned proteins—Proteins bonded by quinone rings.

test—The hollow exoskeleton of an echinoderm.

unios—A group of freshwater lamellibranchs.

vascular—Plants possessing a vascular system: strands of wood and bast, forming tissues for the conduction of water and of food substances.

Index

ABSOLUTE (radio-active) age of rocks, 26–8, 94
agglomerate, 24
agnathids, 48, **Fig. 22**
algae, 23, 36, 41, 53, 82, 83, 86
Algonkian, 26, 41–2
Allerød oscillation, 89, 90
ammonites, **Fig. 30**, 57, 59, **Fig. 33**, 62, 70, 91, **Fig. 48**
ammonoids, 54, **Fig. 28**, 55, 68, 87
amphibians, 52, 53, 56
anaerobic bacteria, 21, 111
angiosperms, 59, 62
annelids, 32, 35
Archaean, 26, 27
archaeocyathids, 36, **Fig. 12**
Archaeozoic, 26
assemblages, life and death, 80
—, reef, 82, 83, 86
axes of contemporaneous movement, 74–6, **Fig. 40**

BACTERIA, 23
ballstones, 46, **Fig. 20**
Beaufort Series, 56, 80
bedding planes, 12, 15, 32, 37
— —, current or false, 17, **Fig. 5**, 83, 88
— —, true, 16, 17, **Fig. 5**
beetles, 90
belemnites, **Fig. 30**, 57, 59, 62, 92
belemnoids, 55
bennettitales, 39, 58, **Fig. 32**
benthonic, 68, 70, 71, 72
biometric studies, 96, 99
biotope, 87, 111
blastoids, 55, **Fig. 46**
Boreal, **Fig. 48**, 92
boulder clay, 88
boundary problems, 72
brachiopods, **Fig. 13**, 42, 43, **Fig. 16**, **Fig. 18**, 46, **Fig. 21**, 48, 50, **Fig. 25**, **Fig. 28**, **Fig. 31**, 57, 68, 72, 79, **Fig. 42**, 83, **Fig. 46**, 96, 103–5, **Fig. 57**, **Fig. 58**
breccia, 21
Bryozoan Bed, **Fig. 9**, 30
bryozoans, 46, 53, **Fig. 28**, 62, **Fig. 36**, 82, 83, **Fig. 46**, 111

CAENOZOIC, 25
Cambrian System, 26, 28, 42, **Fig. 16**, 108
Carbon^{14} dating, 27
Carboniferous System, 26, 28, 49, **Fig. 24**, **Fig. 25**, **Fig. 26**, 83, 108
casts, 31
cave deposits, 21
cephalopods, dibranchiate, 35
Chalk, 22, 25, 59, **Fig. 33**, 107
chondrichthyes (sharks), 48, **Fig. 23**
cidaroids, 55, 57
cleavage, 37, **Fig. 13**
coal, 12, 20, 23
— balls, 31
—, bituminous, 23
—, boghead, 23
—, brown, 23, 62
—, cannel, 23
—, humic, 23
— seams, 12, 50
Coal Measures, 51, **Fig. 26**, **Fig. 27**, 80, 94, 98, **Fig. 53**
coccoliths, 22, 23, 59
concretions, 20
condylarths, **Fig. 35**, 99
conglomerate, 21
conifers, 51
conodonts, **Fig. 11**, 68
coprolite, 33
coralline, 62

INDEX

corals, 36, 46, **Fig. 21**, 50, **Fig. 25**, 55, **Fig. 31**, 57, 82, 83, 87, **Fig. 46**, 90
cotylosaurs, **Fig. 29**, 56
Cretaceous System, 25, 28, 59, **Fig. 33**, **Fig. 34**, 93, **Fig. 49**, 103–5, 107
crinoids, 43, 46, **Fig. 28**, 55, **Fig. 30**, 57, **Fig. 33**, 68, **Fig. 46**, 87
crossopterygii, 49, **Fig. 23**, 52
cycads, 39, 58
cyclothem, 51, **Fig. 27**, 98
cystoids, 43, **Fig. 18**, **Fig. 46**

DEVONIAN SYSTEM, 26, 28, 47, 82, 87, 108
diachronism, 70, **Fig. 38**
diagenesis, 20, 83
diatoms, 23
dinosaurs, 36, 58, 59, **Fig. 34**
dip, 15, **Fig. 6**
distorted fossil, 36, **Fig. 13**
dolomitization, 21, 83, 84

ECHINODERMS, 43, 83, 86
echinoids, **Fig. 11**, 43, **Fig. 31**, 57, 59, **Fig. 33**, 82, 83
elasmobranchs, **Fig. 23**, **Fig. 26**
elephants, 82
Eocene System, 25, 26, 62, **Fig. 36**, 82, 83, **Fig. 47**, 99, **Fig. 56**
epoch, 67
era, 25, 67
evolution, 94–105

FACIES, 46, **Fig. 21**, 71–2
faults, 17, **Fig. 8**
films, 32, 37
flint, 20
folds, 17, **Fig. 7**
—, anticline, **Fig. 7**
—, isoclinal, **Fig. 39**
—, overfold, **Fig. 7**
—, recumbent, **Fig. 7**
—, syncline, **Fig. 7**
footprints, 32
foraminifera, 22, **Fig. 11**, **Fig. 28**, 55, 59, 62, 82
formation, 65, 66
fossils, pyritized, 21, 31, 38
—, silicified, 21
frost heave, 88, 111

GASTROLITH, 33

gastropods, 35, 43, 62, **Fig. 46**, 79, **Fig. 42**, 83
Gault clay, 59, **Fig. 33**, 70–2, **Fig. 38**, 78
geosyncline, 43, 46, 62
glacial, 90
glaciation, 62, 81, 92, 93
Glossopteris flora, **Fig. 29**, 56
Gondwanaland, 93
goniatites, 48, 51, **Fig. 26**, 68
graptolites, 37–8, **Fig. 14**, 42, **Fig. 17**, **Fig. 19**, 46, 68, 72–4, **Fig. 43**
greywacke, 73
gymnosperms, 58, **Fig. 33**, 59

HOLOCENE System, 25, 27, 28, 64, 89
holothurians, 32
horizontal variation, 103
horses, 39, 99–103, **Fig. 54**, **Fig. 55**, **Fig. 56**

ICHTHYOSAURS, 34, 57, 59, **Fig. 34**
impressions, 32
insects, 29, 51, 90
interglacial, 62, 88, 90
interstadial, 88, 90

JELLY-FISH, 32
joint, 18
Jurassic System 25, 28, 57, **Fig. 30**, **Fig. 31**, 82, 83, 91, 92, **Fig. 48**, 103–5, **Fig. 58**, 107

KARROO System, 55, **Fig. 29**

LAMELLIBRANCHS, 31, **Fig. 11**, 35, 51, **Fig. 26**, **Fig. 27**, **Fig. 30**, **Fig. 31**, 57, 59, **Fig. 33**, 62, **Fig. 36**, 70, 79, **Fig. 42**, 82, 83, 86, 98, **Fig. 53**
Law of Superposition, 14, 18, 25
lignite, 23, 62
limestone, 14, **Fig. 4**, 20, **Fig. 9**, 22
—, bioclastic, **Fig. 9**
—, biohermal, 22, 30, 82
—, Carboniferous, 49, **Fig. 25**, 83
—, detrital, 84
—, Magnesium, 53, 83
—, nummulitic, **Fig. 9**
—, oolitic, 22, **Fig. 9**, 74–6
—, stromatolitic, 40
—, Wenlock, 46
lineage, 96
lung fish (dipnoans), 49, **Fig. 23**, 53, 80
lycopods, 49, **Fig. 24**, 51

INDEX

MAMMALS, 58, 60, **Fig. 35**, 82, 99
mammoth, 29, 34, 90
marine band, 51, **Fig. 27**
— —, regression, 58
— —, transgression, 42, 59
Mesozoic, 25, 39, **Fig. 32**, 81, 90, 93, 107
microfossils, 33, **Fig. 11**
micropalaeontology, 33, 34, 65, 77
migration routes, **Fig. 58**
Millstone Grit, 51, **Fig. 26**
Miocene System, 25, 28, 40, 60, 100–1, **Fig. 56**, 106
Mississippian System, 26, 28
molluscs, 82, **Fig. 46**, 90
moulds, 31, 83
mudstone, 12, 22
mussel bands, 80, 98, **Fig. 53**

NAUTILOIDS, 43, **Fig. 18**
New Red Sandstone, 52
nodule bed, 81
non-sequence, 12, **Fig. 4**, 23

OIL, 23, 33, **Fig. 10**, 76–7, **Fig. 41**, 87
Old Red Sandstone, 47–9, **Fig. 22**, **Fig. 23**, 70
Oligocene System, 25, 28, 60, 82, **Fig. 47**, 100, **Fig. 56**, 106
Ordovician System, 26, 28, 42–3, **Fig. 17**, **Fig. 18**, 72, 108
orogenesis, 11, 46, 47, 52
osteichthyes (bony fish), 49, **Fig. 23**
ostracoderms, 48, **Fig. 22**
ostracods, **Fig. 11**, 48, 50, 79, **Fig. 42**, 90
otolith, **Fig. 11**
outcrop, 16, 66
oysters, 57, 83, 84, 96–8, **Fig. 51**, **Fig. 52**

PALAEOCENE System, 25, 60, 82 **Fig. 47**, 99
palaeoclimate, 88–94
palaeoecology, 79, **Fig. 42**
palaeomagnetism, 93
palaeotemperature, 92
Palaeozoic, 25, 26, 46, 81, 83, 94, 108
palynology, 33, 34, 89, 90
pareisaurs, **Fig. 29**, 80
pelagic, 87
pelmatozoans, 43
Pennsylvanian System, 26, 28
periglacial, 88
Permian System, 26, 28, 52, **Fig. 28**, 55, 84–6, 108

period, 25, 67
petrifactions, 30
Phanerozoic, 26, 105, 106
phytoplankton, 87
placoderms, 48, **Fig. 22**
plant communities, 89
Pleistocene System, 25, 27, 28, 30, 64, 82, 89, 90, **Fig. 47**, **Fig. 56**, 106
plesiosaurs, 57, 59, **Fig. 34**
plexus, 98
Pliocene System, 25, 28, 62, **Fig. 47**, **Fig. 56**, 102, 103, 106
pollen, 34, 89
Pre-Cambrian, 26, 27, 40, **Fig. 15**
Primary Era, 25
Proterozoic, 26
provinces, climatic, 91, **Fig. 48**
—, faunal, **Fig. 58**, 105
psilophytes, 51
pteridophytes, 49, **Fig. 24**
pteridosperms, 51, 94
pterosaurs, 57, 59, **Fig. 34**
pyroclasts, 19, 24

QUATERNARY Era, 25, 106

RADIOLARIA, 23, **Fig. 11**, 87
reefs, 46, 53, 82–7, **Fig. 44**, **Fig. 45**, **Fig. 46**, 91, **Fig. 50**
reptiles, 57, **Fig. 34**, 59
rhinoceroses, 82
rocks, arenaceous, 19, 21
—, argillaceous, 19, 21
—, calcareous, 19, 22
—, carbonaceous, 19
—, clastic, 19
—, evaporite, 19, 54, 88, 93
—, ferriferous, 19
—, igneous, 11
—, metamorphic, 11
—, phosphatic, 19
—, rudaceous, 19, 21
—, sedimentary, 11, 12, 18–23
—, siliceous, 19

SANDSTONE, 12, 20
scalecodont, **Fig. 11**
scorpions, 39
sea urchins, 31, 65, 68, **Fig. 37**
Secondary Era, 25
series, 67
shale, 12, 22

shelf seas, 46, 47
silicification, 21
Silurian System, 26, 28, 43, **Fig. 20**, **Fig. 21**, 86, 87, **Fig. 50**, 108
species, 95–105
sponges, **Fig. 11**, 36, 58, **Fig. 33**, 82, 83, **Fig. 46**, 86, 87
spores, 34, 89
stage, 66, 67
stratification, 12, **Fig. 1**
Stratigraphical Table, 25, 27, 28
strike, 15, **Fig. 6**
stromatactis, 83, **Fig. 46**
stromatolites, 41, **Fig. 15**
stromatoporoids, 46, 83, **Fig. 46**, 87
surface, bored, 15, **Fig. 4**, 74
—, striated, 88
synapsids (mammal-like reptiles), **Fig. 29**, 56
system, 25, 76, 72

TAIGA, 64
talus fans, 86
Tertiary Era, 25, 59, **Fig. 35**, **Fig. 36**, 62, 81, 82, 90, 106
Tethys, 54, 55, 59, 62, 91, **Fig. 48**, 92

thecodonts, **Fig. 29**
Theory of Colonies, 73
— — Continental Drift, 93
tillite, 88, 93
time-stratigraphic unit, 67
trace fossils, 32
trails, 86
Triassic System, 25, 28, 40, 52, **Fig. 28**, 55, 107
trilobites, 42, **Fig. 16**, 43, **Fig. 18**, 46, **Fig. 48**, 55, 72, **Fig. 46**, 87, 99
turbidity currents, 43

UNCONFORMITY, 12, **Fig. 3**, 18, 23, 72
Upper Greensand, **Fig. 33**, 70-2, **Fig. 38**

VASCULAR plants, 49
volcanoes, 24

WEICHSELIAN, 89
Woolwich Shell Bed, 80
Würm glaciation, 89, 90

ZONE, 67–72
zone fossil, 68